# THE INVENTION OF INFINITY

# THE INVENTION OF INFINITY

## Mathematics and Art
## in the Renaissance

### J. V. FIELD

*Department of History of Art*
*Birkbeck College, University of London*

OXFORD   NEW YORK   TOKYO
OXFORD UNIVERSITY PRESS
1997

y Press, Great Clarendon Street, Oxford OX2 6DP

Oxford    New York

and    Bangkok    Bogota    Bombay    Buenos Aires
Calcutta    Cape Town    Dar es Salaam    Delhi    Florence    Hong Kong
Istanbul    Karachi    Kuala Lumpur    Madras    Madrid    Melbourne
Mexico City    Nairobi    Paris    Singapore    Taipei    Tokyo    Toronto
and associated companies in
Berlin    Ibadan

Oxford is a trade mark of Oxford University Press

Published in the United States
by Oxford University Press Inc., New York

© J. V. Field, 1997

A catalogue record for this book is available
from the British Library

Library of Congress Cataloging in Publication Data
(Data available)

ISBN 0 19 852394 7

Typeset by EXPO Holdings, Malaysia

Printed and bound in Great Britain by
Butler & Tanner Ltd, Frome and London

To

Arthur Beer
(1900–1980)

# PREFACE

This book grew out of a series of lectures I gave at Imperial College, University of London, while I was an academic visitor in the Mathematics Department. I am grateful to the College for its hospitality, and to several members of the Mathematics Department, most notably Professor Eduardo Ortiz, for their kindly support for my work.

The book grew directly from the lectures, but the lectures themselves had grown from a series of papers and books dealing with what I increasingly saw as various related aspects of the history of the mathematical sciences. Isolating the specific relationship of mathematics with art reflects a long-standing habit of enjoying looking at pictures, and the inevitable acquisition, in the process, of a certain amount of knowledge of the history of art.

I am grateful to the Leverhulme Foundation for a grant which has made it possible for me to write this book.

I should also like to thank the following: Professor Martin Kemp, who made valuable comments on an earlier draft of this work; my colleagues at Birkbeck College, particularly Dr Francis Ames-Lewis, who read and commented upon earlier drafts of Chapters 1, 2 and 6; and Maureen O'Rourke and Dr M. E. W. Williams, who made helpful specific and general comments on the text as a whole.

*London*
December 1996
                                                                 J. V. F.

# CONTENTS

# FIGURE ACKNOWLEDGEMENTS

# INTRODUCTION

The name Renaissance is really a label for a style not for a period. In art and literature its beginnings are usually traced back to the powerful works of the painter Giotto di Bondone (*c.* 1266–1337) and the poet Dante Alighieri (1265–1321). In mathematics and the mathematical sciences the process of 'rebirth' is generally seen as starting later, perhaps in the mid-fifteenth century, with astronomers' attempts to return directly to Greek sources. This scientific Renaissance, like that of the arts, has no definite end—which is to say that books with titles of the form *X and the end of the Renaissance* are a good way for certain kinds of historian to make bids for the attention of their professional colleagues.

The present book, which makes no claim to be a complete history of Renaissance art or of Renaissance mathematics, will look at the relations between the two in the period from about 1300 to about 1650. The first date is that in which Dante set the action of his *Divine Comedy*. The second is a compromise between the date of the death of Galileo Galilei (1564–1642) and those of the inventors of two new, un-Greek, ways of doing geometry: the algebraic geometry of René Descartes (1596–1650), published in 1637, and the projective geometry of Girard Desargues (1591–1661), published in 1639. It would not be reasonable to make the cut-off a sharp one, so we shall in fact sketch in a little of the sequel up to the early eighteenth century. We are thus considering a historical period which does not have a specific name in either history of art or in history of science. The reason for this is not difficult to find: we are concerning ourselves here with interactions between activities that have, traditionally, been treated separately. In fact they have also been treated unequally. There are many general histories of Renaissance art, and many detailed monographs on particular aspects of it. History of science (in which I include history of mathematics) is a much newer discipline than history of art, and its literature is much less rich—though equally uneven. Moreover, given the nature of academic specialisation in the present century, it is understandable that in almost all cases any connections made between the histories, of art and of science, seem to have been too tentative to lead to alterations in the time base of the study in question. The time base of the present study is dictated by its concern with the specific link between science and art that is provided by mathematics.

The period from 1300 to 1650 is a significant one in the history of mathematics because it shows the subject gradually taking up a much more important position in the intellectual map. In its own way, this development is a contribution to the beginnings of a recognisable 'modernity' as striking as many other aspects of the Renaissance. We are indebted to the Renaissance for the characteristically mathematical nature of what we now call science in much the same way that we are indebted to it for the classical architecture of many public buildings and the Roman-style typefaces that dominate our printing. On the other hand, many of the institutions that one may regard as typical of modern European society, and indeed of that of much of the rest of the world, are of pre-Renaissance origin. For instance, hospitals, universities and banks are all characteristically medieval developments.

In any historical work, and irrespective of the historical period concerned, current fashion and the historian's character will play their parts in determining the relative emphasis given to continuity and change. The choice of subject also plays a part. The story told here is largely a social history of mathematics. It is not, of course, the complete history of mathematics in all its social settings (if such a thing could ever be contemplated), but rather a story of how mathematics became important in the everyday lives of more and more people—craftsmen, tradesmen and intellectuals. It is thus a story of slow evolution rather than of sharp revolution, though the texts and the works of art which appear as archaeological exhibits of the social change along the way do, often enough, seem in themselves highly original. They have indeed been chosen for just that quality, but this story is, all the same, intended as the story of the miles, not the story of the milestones.

We shall be concerned largely with the history of the emergence of perspective studies from the world of artists into that of mathematicians. The artistic practice of the fifteenth century led to the invention of a new kind of geometry in the seventeenth. This is undoubtedly the most important of the contributions that art made to mathematics. The new projective geometry of Desargues was to prove to be a highly significant contribution to the development of mathematical ideas. However, the study of perspective by apprentice artists was only a small part of an ongoing tradition of 'practical mathematics'. The pupils of fourteenth- and fifteenth-century 'abacus schools', who included prospective merchants as well as prospective craftsmen, were taught mainly arithmetic, some algebra, and a little geometry, all of it by means of series of worked examples relating to commerce. This syllabus was very different from that of the mathematical sciences taught in universities at the time, which was largely based on very abstract arithmetic and geometry and their application in the theory of music and elementary astronomy. The story told here is partly that of the eventual coming together of the two traditions, the learned and the practical, exemplified by the university and the abacus school, or the scholar and the craftsman. Our chief concern will be with how the practical mathematicians changed the scholarly mathematicians' notions of geometry. There is, however, another component to the interaction of the traditions, namely how the scholarly mathematicians came to take an interest in 'commercial arithmetic', viz. algebra, and went on to develop it into something of equal standing with geometry. This element in the history of mathematics has been well covered in earlier histories. We shall have occasion to mention it here, but the main emphasis will be on the kind of mathematics directly connected with the practice of artists. For this we shall look both at elementary schoolbooks and at learned works written by professional mathematicians.

Much of the content of textbooks of practical mathematics is tediously repetitive, and sufficiently unlike what one now learns at school to be, in places, quite awkard to follow. The present author admits to having felt thoroughly humiliated by the difficulty of making sense of parts of some texts that were clearly in principle very elementary. On the other hand, treatises that propose to advance understanding among learned mathematicians are not necessarily easy for anyone to follow. Their primary concern is often with intellectual rigour rather than clarity, and they are written by professionals for professionals. Some of these texts do turn out to be blessedly lucid, but most do not. We shall accordingly try to explain the principles that are used by particular authors, practical and learned, rather than giving full details of the contents of particular mathematical treatises. All the same, where possible we have included extracts from the works concerned, so as to give some idea of their style. The style is indeed part of the story, since it gives one an idea of the readership the authors in question believed they were addressing. To take two extreme cases: the texts give one a vivid sense of listening to Piero della Francesca instructing an apprentice in the fifteenth century, and some uncomfortable inkling of the headaches most of his mathematical acquaintances got from attempting to understand Desargues in the seventeenth.

In any period, mathematics seems to share with art the capacity for generating works that can be enjoyed irrespective of their historical context. A little thought shows, however, that a 'great period' of the one is not usually so for the other. In general, the art seems to be in advance of the mathematics—in the sense that, for instance, what are now the most admired examples of the art of ancient Greece almost all date from before the most admired mathematics. In this case, the disjunction is a matter of centuries. Things moved faster in Western Europe during the Renaissance: the technical triumph of solving cubic equations, by Scipione dal Ferro (1465–1526), occurred only a little under a century after the technical triumph of Masaccio (1401–*c.* 1428) in giving a convincing illusion of the third dimension in a painting. In fact, the pattern of the evolution of the relations between mathematics and art in the period from 1300 to 1650 definitely tends to have the more famous artists in its earlier part and the more famous mathematicians in its later one. That is, of course, to some degree a function of today's taste in the arts. Since one famously cannot expect to get anywhere by arguing about taste, the imbalance may be regarded as possibly inherent, for some historical reason yet to be adequately explained, but also (more cheerfully) as a good reason for having a more careful look at fifteenth-century mathematics. Looking at geometry rather than algebra does indeed suggest that historians may have underestimated the achievements of some fifteenth-century mathematicians.

One aspect of artists' mathematics will largely be ignored here, namely the consideration of 'proportion'. All through the period with which we shall be concerned, and for long before and after it, proportion was something people talked about. Theoreticians of one kind and another perceived correspondences between the proportions found in music, that is in music theory, which was accepted as derived from the learned science of arithmetic, and the proportions found in, say, buildings or paintings. Sometimes, however, 'proportion' seems to be derived from geometry rather than arithmetic. That is, the measurements concerned are taken from a geometrical figure: lengths of the side and diagonal of a square, say, or the side and height of an equilateral triangle. Sometimes 'proportion' has other more complicated meanings. In short, we find the word 'proportion' used in a number of senses that seem decidedly disparate to a twentieth-century mathematician. The notion of proportion clearly had a definite significance for artisans, in the sense that it was a source of series of related measures. However, the mathematical sources of this useful body of lore are too disparate for one to construct a coherent story. Essentially, we are looking at a mass of fragments taken from Euclid's *Elements*. Parts of the story of that immensely influential work will, of course, be included in our history.

# Chapter 1

## MEDIEVAL MATHEMATICS AND OPTICS AND THE RENAISSANCE STYLE IN ART

Since universities were the recognised custodians of true mathematical learning in 1300—and still by and large retain that status today—it is with this characteristically medieval institution that we shall begin.

### The learned tradition

University teaching was in Latin, Latin being the language of international communication. In fact, the name university comes from *universitas*, meaning 'a universe', that is a copy of the world, consisting of people from all nations (though not of both sexes). Students were sometimes divided into 'nations' for administrative purposes.

The students had learned Latin, and learned to study other subjects in Latin, either privately from tutors or by attending a grammar school. The subjects in which one could obtain a university degree were theology, law and medicine. Courses, and degree subjects, varied from one university to another, as did constitutional matters such as arrangements for paying teachers. Some universities were largely self-regulating, others were sternly under central political control. For instance, the University of Padua, founded in 1202 (and numbering among its first intake of undergraduates Albertus Magnus (*c.* 1193–1280)), was set up as the University of the Republic of Venice and, in characteristically Venetian fashion, was controlled by being entirely state financed. In fact, during the Renaissance, quite a number of towns or states set up universities, apparently as status symbols. Some of these proved to be short-lived.

The basis of university education was instruction in the arts. First came the three arts of the *trivium* (from which we get our word 'trivial'), namely grammar, logic and rhetoric. There followed the four mathematical arts of the *quadrivium*, namely arithmetic, geometry, music and astronomy. Medieval and Renaissance illustrations of these arts are to be found in various contexts as representations of learning in general. See Fig. 1.1.

However, the level to which the subjects were studied was not as high as their learned names may imply to a twentieth-century reader. Grammar meant learning to write, and to spell correctly since, though spelling was highly variable in vernacular languages, variation was not permitted in Latin. Logic largely meant putting words together to make sentences, that is, what we should now call syntax. Rhetoric meant constructing an argument in accordance with logical rules.

The arts of the *quadrivium* were considered more difficult. Arithmetic involved learning to write numbers, to identify and use various arithmetical processes, such as addition and subtraction, and to give their proper names to various kinds of ratios as well as learning to manipulate them in various

**Fig. 1.1** Andrea di Buonaiuto, the four mathematical arts, detail from 'The triumph of St Thomas Aquinas', fresco, *c.* 1365, Cappellone degli Spagnoli, Santa Maria Novella, Florence. The four female figures are personifications, and each has a famous practitioner at her feet. From the left we have Arithmetic (with Pythagoras), Geometry (with Euclid), Astronomy (with Ptolemy), and Music (with Tubal Cain).

ways. For instance, one learned that a ratio in which the first number was one larger than the second was called 'superparticulate', and that the ratio 3:2 was called 'sesquialterate'. Such names continued in use throughout the period with which we are concerned and provide a continual hazard to historians of mathematics lacking a properly medieval education. The recognised practical application of arithmetic was in music, that is in the theory of music, where the arithmetical ratios appeared as ratios of string lengths. For instance the ratio 1:2 corresponds to the interval of an octave between the corresponding musical notes, that is the notes produced by strings whose lengths are in this ratio.

Geometry was studied from Euclid, but that is not to say that the student was expected to read, or have read out to him, all the books of Euclid's *Elements*. At most the courses seem to have got as far as Book 6. That is, the student was taught about triangles—Pythagoras' Theorem is proved in *Elements* Book 1, Proposition 47—something about circles and inscribing polygons in or around them (Books 3 and 4) and something about ratios and proportions, considered geometrically (Books 5 and 6). All this geometry is strictly two dimensional. Solid geometry appears only from *Elements* 11 onwards. The recognised practical application of geometry was in astronomy, which was generally taught from the *Sphere* of Sacrobosco (written in the thirteenth century) or, in more advanced classes, from a Latin version of the first two books of the *Almagest* of Claudius Ptolemy (*fl.* AD 129–141). In Ptolemy's work, originally written in Greek in the city of Alexandria (Egypt), some of the mathematics is difficult. However, the first two books are quite straightforward. The first deals with the system of the world. The second is primarily concerned with an account of astronomical appearances, for instance explaining how to find the length of the day on some particular date in the year and at some particular geographical latitude. Sacrobosco's work covers the same material, but is much less concerned with calculation.

The astronomy course also included the study of astrology, that is the study of the effect of heavenly bodies on the Earth. It should be remembered that in this period the facts that the Sun caused the seasons and the Moon the tides were taken as evidence for astrological influence, that is action at a distance. However, the chief reason for university study of astrology seems to have been its applications to medicine. Thus, when Galileo Galilei, as a Professor of Mathematics, taught astronomy at the University of Padua, from 1592 to 1610, he was teaching standard geocentric astronomy and largely addressing himself to medical students who needed to learn some elementary astronomy in order to be able to practise astrology for medical purposes. For instance, certain angles between the directions in which particular planets were seen (so-called 'astrological aspects') were believed to affect particular bodily fluids and thus to influence the patient's health. A relic of such astrological medical beliefs is found in the name for the disease influenza, a name which literally means 'influence', and originally referred to celestial influence.

The cosmological extensions of astronomy, such as enquiry about why the planets should move as they do, or what they are made of, were considered the province not of the mathematician but of the philosopher. His explanations were, on the whole, expected to be qualitative rather than quantitative; for example, that since the motion of the planet Mars showed regular cycles it was clear that the body of the planet was fixed to a sphere which was part of a system of regularly rotating spheres. It was then up to the mathematician, that is the astronomer, to calculate suitable diameters for the spheres. There were, however, parts of natural philosophy where mathematical investigation was considered integral to any attempt to provide an explanation of the workings of nature. Such subjects were called 'mixed sciences'.

# The science of sight

An example of what was considered an acceptable mixture of mathematical and philosophical theory is to be found in the study of what is today called optics. That name comes from the Greek. It was introduced in the sixteenth century, and is a product of the characteristically Renaissance determination to return explicitly to the Greek origins of the science. The medieval name is the Latin *perspectiva*.

The medieval science of *perspectiva* was a complete science of vision. It dealt not only with the nature and behaviour of light but also with such matters as the anatomy and functioning of the human eye. On the whole, in the Middle Ages, as in ancient times, philosophers were agreed that seeing was an active process. That is, the eye sent out beams. The suggestion that the eye merely received light from external objects was indeed put forward, most notably by the Islamic mathematician and natural philosopher Ibn al-Haytham (AD 965–c. 1040, usually known in the West as Alhazen), but received rather little attention. Perhaps this description of seeing as passive was regarded as psychologically unconvincing—though, as we now know, it turned out to be correct. It now appears that the active part of seeing is not in the eye but in the way the brain processes the information it receives.

In any case, whether sight is by the active emission of eye beams or the passive reception of light, we can still use the same geometrical method to describe certain elements in what we see, since eye beams, like light, are assumed to travel in straight lines. Thus Euclid's *Optics* is written in the same style as his *Elements*, that is as a succession of mathematical theorems without linking explanatory text. It completely ignores problems of physics and sets about proving what can be proved by means of geometry, for instance that if two magnitudes are of equal size then the one which is nearer the eye will appear larger—in the sense that it subtends a larger angle at the eye (see Fig. 1.2).

## Naturalism in art

Euclid's *Optics*, written in Greek in the third century BC, was well known in Latin translation throughout the Middle Ages. This is obvious to anyone who studies the works of philosophers of the time. It is not obvious at all if one looks instead at the works of artists. These regularly flout the mathematical rules that Euclid shows as establishing the relative sizes that subjects will present to the eye. For instance, in scenes of the Last Judgement it is normal, as in the mosaic shown in Fig. 1.3 (detail in Fig. 1.4), to make Christ very large, and to make other figures smaller in proportion to their spiritual or narrative significance. If the artists responsible for such scenes asked or received advice from the learned about their work it was clearly from theologians rather than natural philosophers. Indeed, everyone would have been agreed that the function of art was not naturalistic representation, but rather the expression of spiritual power. Even in scenes where all the figures are presented as human rather than divine, we find the same theological realism, portraying importance through size, rather than an attempt at scientific naturalism.

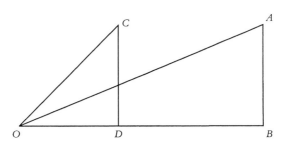

**Fig. 1.2**   The eye is at the point *O*. The two magnitudes *AB* and *CD* are equal, but *CD* is nearer the eye, so ∠*COD* is greater than ∠*AOB*. This theorem is proved in Euclid's *Optics*.

**Fig. 1.3**  Mosaic from the baptistery of S. Giovanni (Florence), *Last Judgement*.

**Fig. 1.4** Detail from Fig. 1.3 showing Christ in majesty.

A turning towards scientific naturalism begins to appear with some force from the late thir-teenth century onwards. This has its theological counterpart, or perhaps theological motivation, in the teaching of Francis of Assisi (*c.* 1182–1226, canonised in 1228). St Francis saw study of the natural world as leading the soul to God, since God was its creator and ruler.

The most striking examples of this newly naturalistic style are in the work of Giotto di Bondone. He was, in fact, one of the most talented story-tellers of all time. Something of the vivid comic-strip character of his art can be seen in the three scenes from the story of Joachim and Anna (the parents of the Virgin Mary) shown in Figs 1.5, 1.6 and 1.7, which come from a cycle of frescos in the Arena Chapel in Padua. In the picture shown in Fig. 1.5, Joachim, ashamed of being childless, has withdrawn to live among his shepherds. He avoids meeting their eyes; they glance at one another questioningly. The human and animal figures are all to scale (sheep at this time probably were scrawny), but the landscape has been shown rather small. The trees have detailed leaves, so do not look as if they were distant. It is clear that Giotto is not constructing a landscape, he is providing a landscape setting for the people in his story. The same landscape appears in a later scene, shown in Fig. 1.6, where Joachim, locked in sleep, dreams of an angel that comes to tell him his wife is pregnant. Though there is no question of an actual interaction between the figures, Giotto has given the angel a movement which is clearly directed towards the sleeping Joachim. A little later in the story we have the scene shown in Fig. 1.7, where Joachim and Anna meet beside the Golden Gate of Jerusalem. As in the earlier scenes, the human figures are all to scale, the embracing figures of husband and wife being emphasised not only by their

**Fig. 1.5** Giotto, *Joachim comes to the shepherds' hut*. Fresco, Arena chapel, Padua.

**Fig. 1.6** Giotto, *Joachim's dream*. Fresco, Arena chapel, Padua.

**Fig. 1.7**   Giotto, *Joachim and Anna meet at the Golden Gate*. Fresco, Arena chapel, Padua.

haloes but by their forward position and the strong vertical axis of the gate tower behind them. As for the landscape in the earlier scenes, the castellated gate is there as scene setting. To show it properly, that is to make it readable as a city gate, Giotto has had to make it far too small in relation to the figures. Simple naturalism has taken second place to the demands of story telling. The gate is merely a secondary element; what the picture is about is the loving embrace of Joachim and Anna.

Similar concentration of purpose is evident in many more of the frescos of the cycle to which these scenes belong. For instance, in a much more prominent position on the wall of the chapel, we have the slightly larger and much more crowded scene in which Judas betrays Christ (Fig. 1.8). Here the embrace is very different from that of Joachim and Anna. Christ has made no move. He meets Judas's eye with dignity, but His figure is all but lost under the vivid yellow of Judas's enveloping cloak. The picture is crowded with figures, but the composition allows no doubt as to what is the important part. In the scene of the lamentation over the dead Christ, shown in Fig. 1.9, the additional figures participate more directly. The landscape setting consists of little more than a wedge-shaped sharply ridged piece of rock and a single tree. The wedge shape, which emphasises the similar shape made by the arms of St John, thrown back in a huge gesture of grief, drives us towards the bottom left of the picture, where the other human figures surround and support the dead body of Christ. In the sky, which is of the vivid (and very expensive) lapis lazuli blue that characterises the whole fresco cycle, small angels are throwing themselves around in

**Fig. 1.8**  *The Betrayal of Christ*. Fresco, Arena chapel, Padua.

**Fig. 1.9**  *Lamentation over the dead Christ*. Fresco, Arena chapel, Padua.

attitudes of violent mourning, in fact keening over the dead as mourners still do in many countries of Southern Europe. If angels may be said to be shown naturalistically, then these are naturalistic angels.

## Giotto's mathematics

As we have already mentioned, the pictures by Giotto shown in Figs 1.5 to 1.9 are all executed in fresco. That is, they were painted into wet plaster. The pigment reacts with the wet plaster—usually changing colour as it does so—and the picture is henceforth an integral part of the surface of the wall. Alterations can be made only by removing the offending part of the plaster, with a chisel, and inserting a new patch, which will inevitably show up as a patch. This is to say that fresco paintings had to be carefully planned. Thanks to the survival of instructional texts, we have a fairly detailed record of the normal procedures.

First, the wall was covered with a layer of coarse plaster. The finer plaster was then laid on in sections, starting at the top of the wall, and usually at the left, each section being what was to be painted that day. In fact, the plaster had to dry for about half an hour after application (the time would be longer if the weather were cold or wet), then the painter could put his colours onto it for about an hour (perhaps a little longer, depending again on the weather) before it became too dry. After that the plaster was left to dry for about 24 hours, before the next section of wet plaster was put in place to be painted. Each section of the plaster is called a *giornata*, that is a 'day's-worth'. The size of each *giornata* would depend on how detailed that part of the picture was to be, and on how fast the painter worked. Generally, by gently running a finger over the surface, one can feel where the plaster of successive *giornate* overlaps, so it is usually possible to trace the painter's progress in some detail.

Giotto's fresco cycle in the Arena Chapel is, like many fresco cycles of its time, actually arranged in rather the same way as a comic strip. Care has also been taken to draw attention to significant non-temporal relationships between certain scenes, such as those of messages brought by angels, but one essentially reads the story scene by scene along rows of pictures, separated from one another by framing elements that are painted to look like strips of white marble with repeated patterns of coloured inlay. Clearly, the positions of these framing elements, and probably some elements of the pictures themselves, would have to be marked on the coarse plaster before the finer plaster was applied. In particular, the 'marble' frames would probably be painted by assistants and apprentices, no doubt using some kind of template to ensure that the repeated patterns were repeated accurately. In fact, getting the 'frames' onto the wall correctly was effectively a surveying task, done on a vertical surface, and clearly by no means trivial. The mathematics involved is simple, but the necessity for accuracy in the procedures is self-evident and no doubt imposed a considerable discipline.

Moreover, the plaster itself and the pigments posed another mathematical problem. The ingredients would have been supplied by an appropriate dealer, probably an apothecary in the case of the pigments. However, everything was natural in origin and therefore (in our terms) liable to be highly variable in its chemical composition. That is, one batch of plaster might react differently with different batches of as-near-as-possible the same pigment, and different batches of plaster might react differently with pigments from the same batch. So mixing batches might give different colours. Actually, it is amazingly rare for the edges of *giornate* to be made visible by unwanted changes in colour. Painters presumably took good care to check how much they needed of the ingredients of the plaster and the various pigments. Blue was a particularly awkward colour. The best blue was ground up lapis lazuli, imported from the East, and very expensive. One had to be careful not to grind it too fine, or the colour was spoiled. Moreover, the blue had to be put on in a separate process, after the wall had dried, using a mixture of egg and water as a vehicle for the pigment. This technique is called working *a secco* ('dry'). The vehicle for the colours applied to the fresh plaster was pure water. Thus the lavish use of blue in the Arena Chapel is intended as an extravagant gesture, and Giotto's patrons no doubt checked carefully to see that they were getting good value for money. Gold was generally used sparingly on fresco, because it tended to darken or peel off and need to be renewed. It had to be put on in the form of gold leaf, held in position by glue, the surface having been roughened a little to hold the glue. The areas for the use of gold and of particular pigments are sometimes specifed in contracts drawn up between Renaissance patrons and artists, but generally the most detailed clauses refer to the use of blue and the sizes of the figures of the patron and his or her family.

In any case, when one looks at the matter a little closely, it is clear that there is quite a lot of mathematical skill required in setting up and painting a fresco cycle, even if the style of the pictures does not in itself seem particularly mathematical.

## Useful mathematics

Viewed simply as mathematics, the surveyor's geometry and craftsman's arithmetic to which we have referred are not particularly interesting. What is interesting about them is that at least from the late thirteenth century onwards such mathematical skills were recognised as useful in wider contexts and were increasingly taught in abacus schools specially set up for the purpose. These abacus schools did their teaching in the vernacular. Sometimes they were set up by individual teachers, sometimes by craft guilds. Sometimes, as in Arezzo (Tuscany), the city council paid a salary to an 'abacus master' (*maestro d'abaco*) who was to give tuition to a specified number of pupils. In Florence, one of the best abacus schools, in the late fourteenth century, was that run by the Goldsmiths' Guild.

The story of Archimedes (*c.* 287–212 BC) investigating how much gold had been used to make his cousin King Hieron's crown, and discovering 'Archimedes' Principle' in the process, could almost have been invented as a fable to show the usefulness of mathematics for goldsmiths (and their clients). Moreover, by the late thirteenth century new uses for mathematical skills were appearing: international banking had been invented and other forms of banking had rapidly become considerably more sophisticated. The origins of banking are multiple and complicated, but one may none the less identify a few of the factors at work. One was the wish to make war, say, on a 'fight now and pay later' basis. That is, the potentate concerned needed to raise money in advance of the arrival of his tax revenues (which would probably come in largely at the time of harvest). There are, indeed, records of occasions when fitting out an army and sending it off to beseige the city in question was apparently considered the most effective way of ensuring that the taxes due were in fact paid. Bankers were able to finance such operations. Among the first was the bank set up in the late thirteenth century by the Roman family of the Frescobaldi. Its clients included the Pope. Over succeeding centuries the Frescobaldi became extremely rich, and gradually bought up large tracts of the wine-growing area of Tuscany—as one may check by reading the small print at the bottom of the labels on bottles of Chianti.

In Florence, which was a centre for international trade, and consequently much concerned with establishing the financial viability of merchants by means of letters of credit, the banking families included the Bardi and the Peruzzi. In the early fourteenth century, the Bardi and the Peruzzi families each paid Giotto to decorate a chapel for them in the Franciscan chuch of Santa Croce (the church of the Holy Cross) in Florence. Both banks collapsed when King Edward III of England (1327–1377) reneged on a large debt. This failure precipitated a financial crisis in 1342–1343. In nearby Siena, the oldest bank grew out of the municipal pawn shop and debt office, called the 'Monte'. This bank too dates from the end of the thirteenth century. For professing Christians there were perceived to be religious difficulties about making a profit from handling money—which was why so much of this kind of business was traditionally carried on by Jews—and the newly formed banks got round the problem by being considered as charities. To this day, a large number of Italian banks, including for instance the Monte dei Paschi of Siena, are technically charities and are consequently required by law to give a certain proportion of their profits to charitable uses, such as educational purposes or relief of the poor.

The activities of merchants and of banks clearly stimulated the provision of elementary mathematical education on a wider scale. The business of abacus schools was largely to teach the 'commercial arithmetic' appropriate to such activities. Surviving manuscripts, which generally seem to be the teacher's copy, to be read out loud to his pupils, contain numerous arithmetical problems relating to such matters as the purchase of sugar cones at so much per pound in Alexandria, their transport, at so much a pound, by sea to Messina and then to Pisa, and then over land to Florence and how much must the seller charge if he is to make a profit of one third? Alternatively, we find a man who decides to pay his creditors: to the first he gives half of what he has, to the second one third and to the third one quarter, leaving himself with only one florin, so how much money did he start with? Some of the problems are, however, framed in less everyday terms. For instance, in the *Libro d'abaco* of Leonardo of Pisa (*c.* 1170–*c.* 1240, sometimes called Fibonacci), the work from which abacus books take their name, we have a problem about a lion, a leopard and a bear engaged in eating a single sheep. The lion eats a sheep in four hours, the leopard in five and the bear in six. So how long will it take them all, eating together, to consume a single sheep? The solution begins by considering what will happen in 60 hours (leaving it to the pupil to notice that this is the lowest common multiple of the eating times), and boldly asserts that in 60 hours a lion will eat 15 sheep. (The answer is that all three animals, eating together, will finish a sheep in $1\frac{23}{37}$ hours.)

Abacus books generally begin right at the beginning, with instructions on how to write Arabic numerals—introduced to the West in Leonardo of Pisa's *Liber quadratorum* of 1202. There follow problems whose solution is arithmetical. Later come problems whose solution involves handling an explicit 'unknown', called *cosa*, which literally means 'thing' and later earns algebra the Latin title *cossa*. However, even early in the seventeenth century it was usual to regard algebra as merely a part of arithmetic. This was because the answers came out in numerical form, but the usage had a respectable ancient ancestry, as became apparent with the rediscovery of the Greek origins of algebra in the *Arithmetica* of Diophantus of Alexandria (third century AD), first printed in 1572. (It was in the margin of his copy of this book that the distinguished French mathematician Pierre de Fermat (1601–1665) was to write his famous last theorem.)

The algebra, and indeed the arithmetic, of the abacus books was deeply indebted to Islamic sources—though the so-called Arabic numerals had in fact been invented by the Hindus, only the zero being an Arabic contribution. The relatively few geometrical problems that appear, generally at the end of the book, are usually reminiscent of propositions found in Euclid, but are almost always proposed in numerical terms. For instance, where Euclid asks one to draw a circle to touch all three sides of a triangle (*Elements*, Book 4, Proposition 4) the abacus book version is to say that we have a triangle with sides 13, 14 and 15 and want to know the diameter of the largest circle that can be placed within it. This 13, 14, 15 triangle is in fact something of a favourite with the writers of abacus books, since it has a number of simple numerical properties (see Appendix).

## Goldsmiths and others

Among those trained as goldsmiths, and accordingly taught the kind of mathematics we have been discussing, are some of the most famous artists of the fifteenth century. At the time, however, they would have been thought of as craftsmen, though some of the more eminent were highly regarded as well educated and highly skilled craftsmen. There was no such profession as 'artist'. People who used their manual skills to make their living were, in principle, regarded as craftsmen or artisans. Indeed Michelangelo Buonarroti (1475–1564), whose spectacular career was to do much to remove this attitude, recollected that when he told his father, who was a lawyer (a *notaio*, that is a notary), that he wanted to become a sculptor, the reaction was the horrified exclamation 'My son, a stonemason!' Although in the fourteenth century most painters were specialists—in contrast to the pattern we find in the sixteenth century—Giotto was an exception. He was responsible (though perhaps only in the capacity of overseer of the building works) for the construction of 'Giotto's Tower', the bell tower of Florence cathedral. Figure 1.10 shows the group of buildings to which this tower belongs. The oldest part is the Baptistery. Its roof, in the form of an octagonal pyramid, is to be seen on the left. In Giotto's time the cathedral had not yet been enlarged to its present size, but planning decisions had already been taken to ensure that surrounding buildings did not come too close. The rather theatrical effect produced by the group of buildings was intended from the late thirteenth century onwards. The final touch was to be the huge dome of the cathedral, designed and built by Filippo Brunelleschi (1377–1446) but not actually completed until after his death.

Even in his own lifetime, it was the dome of the cathedral that made Brunelleschi famous, but he was, in fact, trained as a goldsmith. His mathematical activity will be discussed in more detail in later chapters. As we shall see, his work included painting and sculpture as well as the design of machinery and buildings. One of the few surviving pieces that relates to his initial training in handling metals is the relief panel he proposed as his entry for the competition to design a new pair of bronze doors for the Baptistery. The competition was not won by Brunelleschi but by another man also trained as a goldsmith, Lorenzo Ghiberti (1377–1455). Naturally enough, this result did not please Brunelleschi. However, the officials of the Wool Guild, which was financing the building of the cathedral, decided that Ghiberti should be Brunelleschi's colleague when Brunelleschi was appointed overseer of the building works at the cathedral. Ghiberti seems in fact to have been more or less a consultant for work on the dome and the east end of the building—his contract required him to spend fewer hours on site than Brunelleschi did.

**Fig. 1.10**   Florence cathedral with the bell tower and the baptistery.

Surviving documentation of their collaboration does not bear out the oft-repeated assertion that the two men were on bad terms with one another. And their work went well. Brunelleschi lived to see his dome all but complete—his dying words were reported to have been 'Do not forget the lantern' (its weight was essential to the stability of the dome)—and Ghiberti went on to finish the doors, and indeed to make another more famous pair (shown in Fig. 1.11). Michelangelo, who was usually very niggardly with praise, described Ghiberti's second pair of doors as being fit to be the doors of Paradise and the name 'Doors of Paradise' has now become stuck to them.

The gracefulness of Ghiberti's work is very much in tune with the majority taste of its time. On the whole, the fifteenth century liked its art to be pretty. The individual panels also display Ghiberti's mathematical skill. Each scene is shown in apparently correct perspective. The first mathematical rule for getting the perspective correct had been invented by Brunelleschi in or shortly before 1413. The problems of checking the actual correctness of perspective pictures will be discussed in later chapters, but one can hardly doubt that Ghiberti has used some mathematical technique to obtain the effect of optical correctness. We know from his writings that he was familiar with many texts that discussed such problems. In any case, it is clear that Ghiberti's treatment of the scenes he shows is essentially much more naturalistic than, say, Giotto's. In particular,

**Fig. 1.11** Ghiberti, Doors of Paradise. Florence, baptistery of the cathedral.

**Fig. 1.12** Ghiberti, panel from Doors of Paradise, Jacob and Esau. Florence, baptistery of the cathedral.

we may note that, in the scene shown in Fig. 1.12, foreground figures are much bigger than background ones and in order to make the buildings completely visible, Ghiberti has had to put them a long way back in space.

Ghiberti has in fact made quite deep spaces in several of his panels. It is perhaps in mitigation of the disturbing effect these multiple spaces might have on the viewer that Ghiberti has not gone all the way with naturalism: the eye heights he assumes for the various scenes do not make it possible for all of the pictures to be viewed correctly from the same position. On the other hand, such mitigation may have been no part of Ghiberti's intention. Unattainable eye heights are to be found in many perspective pictures of Ghiberti's time. Thus the Doors of Paradise are, in fact, rather good examples of the way perspective was used in the fifteenth century.

# Chapter 2

# BUILDING, DRAWING AND 'ARTIFICIAL PERSPECTIVE'

To the twentieth-century eye, correct perspective seems an obvious contribution to making a picture more naturalistic. However, the invention of mathematical rules for this purpose does not seem to have a direct link with the increased naturalism which is a feature of the painting of the fourteenth century. It was not a painter who invented a method of showing objects exactly in accordance with the laws of optics: Brunelleschi was, as we have already noted, trained as a goldsmith, and he was, as far as we know, never a member of the guild that painters belonged to, namely that of the Physicians and Apothecaries (*Medici e Speziali*), or of the painters' confraternity, the Company of St Luke (*Compagnia di San Luca*).

In connection with his invention of 'artificial perspective' (as it was generally called in the fifteenth century) Brunelleschi did make at least two paintings, but he would probably have described himself as being an engineer (*ingegniere*). His training as a goldsmith also taught him how to work with other metals, and very probably also how to make mathematical instruments such as quadrants or astrolabes. The earliest biography of Brunelleschi, finished shortly before his death, tells us that he made clocks. His other known activities included designing stage machinery for the elaborate pageants that took place in the church of Santa Maria del Carmine on the festival of the Ascension. Thus, although Brunelleschi is now best remembered for designing buildings, and overseeing the building works, to call him an architect is very misleading. In fact, the term architect is not used in its modern specialised sense until the sixteenth century. In the fifteenth century, anyone who had a right to pass an opinion on the design of a building, a patron or even perhaps an official with financial rather than aesthetic responsibility for it, could be called an architect.

## Reviving the ancient style

The profession of architect grew out of the deliberate revival of the classical style. This style is described in a single ancient treatise, Vitruvius' *On architecture* (*De architectura*), probably written about 40 BC. Vitruvius gives detailed accounts of building techniques, but what chiefly interested his fifteenth-century readers was his stress upon the importance of proportion. Proportions referred to all elements of a building, linking the sizes of one with another. As a supplement to Vitruvius, one could, of course, study the remains of actual Roman buildings, to be found in many parts of Italy, but most abundantly in Rome itself.

Thus the intellectual materials for reviving the ancient style of buildings would seem to be ready at hand. But actual building was a craft. Generally—and this is true not only in the Middle Ages but also through much of the Renaissance period—a new building was explicitly planned as a development of one that had been built fairly recently. Designs and construction techniques were adapted from those used in the earlier building. Often enough, some of the same workmen must have been employed. Moreover, construction was always a slow process, so that, for instance, the appearance of cracks would give adequate warning that parts of the structure needed to be strengthened. One way of recognising that one is in an old building is the presence of staircases and walkways giving access to all parts of the structure: these allowed the engineer in charge to make regular checks to see where repairs and reinforcements might be needed.

Attempts to revive an ancient style of architecture did not greatly change the actual building techniques that were used, but they did introduce some new elements into the design process. In fact, they were essentially responsible for separating out design as a distinct process—and thus, in the longer term, for making architectural design a profession separate from that of overseer of the building works.

Since Vitruvius gave such importance to proportion, it was clear that mathematical methods were appropriate not only in laying out the ground plan of a building or deciding on its height, but also in such details as the design of column capitals. Now, many details of Renaissance buildings are copied, with varying degrees of exactness, from surviving fragments of ancient buildings. When Brunelleschi made drawings of buildings in Rome, as he is known to have done, he made them as a preliminary to making actual copies of the parts he had drawn, and it is accordingly to be presumed that even if the drawings were not done exactly to scale (as they would probably be done in our own time) they were at least marked with the required proportions.

Brunelleschi's buildings, such as the churches of San Lorenzo and Santo Spirito in Florence, are notable for their display of mathematical proportions. By his placing of columns and arches, Brunelleschi makes sure that one can see the proportions of spaces clearly. For instance, repeated elements are used to make it obvious that the ground plans of certain spaces are square or are rectangles with their sides in some particular ratio such as 4:3. The characteristic use of grey stone—a millstone grit from the hills near Florence, called *macigno*—produces a contrast between white walls and grey columns and mouldings that makes the space eminently readable in simple mathematical terms (see Fig. 2.1).

## Brunelleschi's 'artificial perspective'

It has been suggested that it was a concern with such readability that led Brunelleschi to invent his rule for 'artificial perspective'. For if particular proportions had been built into the structure it was, clearly, a matter of interest for the designer to ask whether the eye, subjecting the building to the rules of 'natural perspective' (that is, the mathematical laws governing natural vision), would see those proportions for what they were. In modern mathematical terms, Brunelleschi would thus be interested in a form of invariance, namely the fact that natural optics left the proportions unchanged. Unfortunately, no writings about perspective by Brunelleschi have survived—and it is, of course, perfectly possible that he in fact never wrote anything on the subject. We do, however, know that he painted two pictures as showpieces of his mathematical rule. Neither has survived.

The first picture showed the Baptistery of Florence cathedral (see Fig. 2.2), as seen from a position just inside one of the doors of the cathedral. The present façade of the cathedral dates from the nineteenth century, but it is still more or less possible, by standing well back in the doorway, to look at the view Brunelleschi painted. It showed the building as seen from a point directly

**Fig. 2.1** Interior of the Old Sacristy, San Lorenzo, Florence, designed by Filippo Brunelleschi.

**Fig. 2.2**  Baptistery of the cathedral, Florence.

opposite the vertical centre line of one of its faces, that is, directly opposite one doorway. As the plan of the Baptistery is octagonal, Brunelleschi had set himself the task of showing a rectangular wall seen straight on and two similarly shaped walls receding at 45 degrees. He had, also, decided to paint a picture of what was at the time regarded as a classical building. The Baptistery is now believed to date from between the eleventh and thirteenth centuries, and its syle would be described as Romanesque. To Brunelleschi and his contemporaries, however, it appeared to embody a local Florentine variant of the classical Roman style, and its windows, arcading and pattern of marble inlay, were taken as exemplars of the kind of simple mathematical proportions Brunelleschi was to use in his own buildings.

We do not know how Brunelleschi's drawing of the Baptistery was done, but we do know that the painting was made on a wooden panel, and that the area above roof level was not painted but silvered. The picture was viewed through a peephole drilled in the panel, the painting being seen reflected in a mirror that was held in the spectator's hand. The silvered area of the panel reflected the real sky, thus enhancing the sense of reality.

The peepshow arrangement for viewing the picture presumably made the three-dimensional effect more impressive, since before setting it up Brunelleschi must certainly have had a chance to notice that a three-dimensional effect was still produced even if the position of the viewer's eye was not controlled in this way. In any case, Brunelleschi's next panel was not designed to be viewed through a peephole. This time, the chief building shown was the Palazzo della Signoria (now called the Palazzo Vecchio) and the viewpoint was close to the opposite corner of the square, on its corner with the street now known as the Via dei Calzaiuoli (see Fig. 2.3). Instead of silvering the sky, as he had done in the picture of the Baptistery, in this second picture Brunelleschi cut away the panel along the line of the top of the buildings, so that the picture could be seen against a

(a)

(b)

**Fig. 2.3** Palazzo Vecchio from the corner of the Via dei Calzaiuoli: (a) modern photo, (b) drawing by Jacques Callot (1592–1635).

real sky. This would, of course, have allowed him to verify the accuracy of the roof line by viewing the picture against the actual scene.

# Alberti's construction (1435)

The view of the Palazzo della Signoria across the square certainly presented a rather more complicated drawing problem than that posed by the view of the Baptistery, but the two panels were apparently painted in fairly rapid succession and in the absence of evidence to the contrary it seems likely that much the same method was used to produce both pictures. Unfortunately, we have no direct evidence what this method was. The first written account of a method of constructing pictures in correct perspective dates from about 20 years later. It is found in the first book of a short treatise *On painting*, written by the learned humanist Leon Battista Alberti (1404–1472).

Alberti's family originated from Florence, but had been exiled, so that Leon Battista first returned to his native city in the 1430s, in the train of Pope Eugenius IV. On his arrival he found that many interesting things had been happening in painting in Florence and, eager to relate them to his own theories about art, decided to write a treatise on the subject. Being addressed to the learned, the first version of the work, finished in 1435, was written in Latin, with the title *De pictura* (*On painting*). A vernacular version (*Della pittura*) appeared in the following year, dedicated to Filippo Brunelleschi. Fifteenth-century conventions did not actually require formal acknowledgement of one's predecessors, so it is not clear whether this dedication is in fact an acknowledgement that Alberti is using Brunelleschi's work or, as it might be in the twentieth century, an elegant method of evading the charge of having appropriated it without due acknowledgement. It is, moreover, possible that Alberti dedicated the work to Brunelleschi merely as a homage, in recognition of the older man's interest in the visual arts.

In any case, Alberti's account of perspective is not addressed to artisans. He is merely sketching the outlines of a procedure for those who wish to understand something about it. The same is in fact true of Alberti's treatment of other aspects of painting. For instance, his long discussion of what he considered suitable subjects for pictures is clearly addressed not to painters, who would have been told what to paint, but to their patrons, who did the telling.

As the first surviving account of a method of perspective construction, Alberti's brief description of how to draw a picture of a chequerboard floor has often been subjected to the kind of detailed, awe-stricken analysis usually reserved for Holy Writ. To a mathematician's eye, however, it is obvious that the description is inadequate. Either Alberti is not explaining himself very well— perhaps finding it difficult to explain mathematics in a way that makes it accessible to a readership whose real interests lie elsewhere—or he himself has not really understood the method concerned. However, since the main outline is clear, it is quite easy for a twentieth-century reader to fill in the gaps in Alberti's account. The construction is shown in Fig. 2.4.

The problem is to construct the perspective image of a square chequerboard pavement with one edge running along the ground line of the picture, that is the line of intersection of the picture plane with the ground. The outer edge of the actual picture is shown as a square centred on the point $C$. This point is called the 'centric point' of the perspective and is the point in the picture directly opposite the viewer's eye, that is the foot of the perpendicular from the eye to the picture plane. The line $ONC$ has been drawn parallel to the ground line, and the distance of the eye from the picture plane is equal to $ON$.

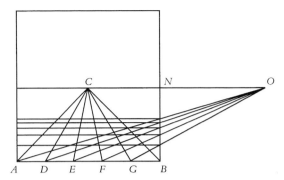

**Fig. 2.4** Alberti's construction (1435).

The method of construction is as follows. First, the lower edge of the picture is divided into the appropriate number of tiles to represent the pavement. Here we have chosen to have five tiles, and the points of division of *AB* are *D, E, F, G*. The second step is to join *A, B* and the points of division to *C*. Alberti somehow fails to make it clear why this is what one has to do—which may perhaps indicate that he expected his readers to know about it already. The lines *AC, DC, EC* and so on represent the lines of the tile edges that run perpendicular to the plane of the picture. The present-day term for such lines is 'orthogonals'. There seems to be no Renaissance equivalent, so for the sake of clarity the modern word will be used. (It is, of course, in general rather dangerous for a historian to depart from period vocabulary, since doing so may tacitly introduce a modern style of thought that distorts one's view of a non-modern line of argument.) The third step of the construction is to join the point *O* to each of the points *A, D, E, F, G*. The fourth step is to draw through the points where *OA, OD, OE, OF* and *OG* intersect the vertical edge of the picture, that is the vertical through *B*, lines parallel to the base line *AB*. These lines represent the lines of the tile edges that run parallel to the ground line of the picture. There seems to be no Renaissance name for them. Their modern name is 'transversals'. The perspective image of our five by five pavement is the trapezoidal grid bounded by the lines *AC, BC, AB* and the uppermost transversal.

Before giving this construction, Alberti has repeatedly referred to the 'cone' or 'pyramid' of vision, that is the sheaf of rays between the eye and the scene that is viewed, and has said that the picture represents a section through this cone. His use of the words cone and pyramid to indicate solid figures whose bases are of irregular shape is a little confusing mathematically, but was quite usual at the time. What is much more serious is that he says not a single word to indicate the nature of the connection between the cone of vision and the construction that follows. Perhaps Alberti expected his learned readers to allow such technicalities to be left to the experts. Although the general level of mathematical education was rising in the fifteenth century, it was not comparable to that prevailing in Western Europe in our own time. Alberti is addressing himself to an upper class readership. On the whole, such readers seem to have known less mathematics than their social inferiors in the artisan classes.

It is, in any case, not very difficult to provide a justification for Alberti's construction. Let us add a point *P* to represent the position of the viewer's feet. In Fig. 2.4, *P* will lie directly beneath *O* on the line *AB* produced. We are using *O* and *P* as the letters for the positions of eye and foot because these are the letters most generally used in perspective treatises of the following century. They are clearly derived from the Italian words for eye and foot, namely 'occhio' and 'piede'. The plan and a vertical section of the set-up are shown in Fig. 2.5. It is clear that, in the diagram of the section, the point in which the line $OH$ cuts the vertical line through $M$ gives the position in which $H$ will appear as seen from $O$. Thus the transversal through $H$, that is the back line of the pavement, will appear in the picture plane as a line parallel to the ground line at the height of this point of intersection. The heights of the nearer transversals will clearly be given by the points of intersection between the vertical through $M$ and the other lines of the pencil radiating from $O$. So we could think of Alberti's construction, as shown in Fig. 2.4, as being put together by superimposing everything in the vertical plane shown in Fig. 2.5(b) on a view of the actual picture, whose significant part is contained in the triangle $CAB$ in Fig. 2.4.

This is not, of course, a proper proof since we have not shown why the orthogonals should converge to $C$. However, there is nothing anachronistic in using superposition. It was regularly used in the Renaissance, as it had been in earlier periods, to show how the façades and internal structures of buildings were related to one another. In the fifteenth century, a learned justification

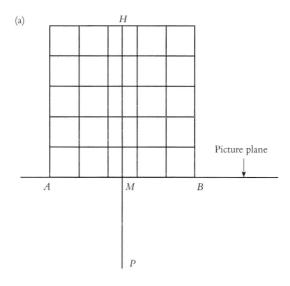

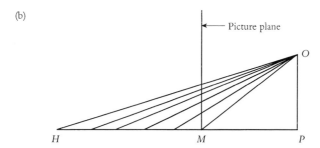

**Fig. 2.5** Plan and vertical section corresponding to the Albertian construction shown in Fig. 2.4: (a) plan and (b) vertical section.

for its use could have been found in the newly recovered text of Vitruvius' *On architecture*, where superposition of lines from perpendicular planes is used in the famous 'analemma' construction for the lines on sundials. Moreover, be it said, Vitruvius explains himself as little as Alberti. As we shall see in Chapter 7, professional mathematicians moved in on Vitruvius' mathematics at about the same time as they moved in on perspective, that is in the second half of the sixteenth century.

The octagonal plan of the Baptistery could, of course, be drawn fairly easily on a chequerboard pavement like that in Alberti's example, using it as a kind of coordinate system. On the other hand, Brunelleschi's second perspective picture seems to have shown an oblique, almost diagonal, view of the Palazzo della Signoria and the open area in front of it, both of which are roughly square in plan. This does not seem to be a very natural progression if the construction used is primarily designed to produce sets of orthogonals and transversals. All the same, both pictures certainly could have been produced by some such method. What we know about them thus unfortunately fails to shed much light on the degree of originality in Alberti's discussion of perspective construction.

Though it cannot have been very prominent in Brunelleschi's panels, the Albertian square-tiled pavement, or more complicated variants of it, are quite common motifs in works of art from the 1430s onwards. They are clearly intended to convey to the viewer that the picture is in correct perspective. Such 'look at this for a piece of mathematics' floors are generally referred to by art historians as *pavimenti*—since most of the treatises that explain how to construct them are written in Italian. Of course, as we have already hinted in connection with Brunelleschi, it is not necessarily easy, or even possible, to work out how the perspective of a picture has been constructed. None the less, the small panel painting of the *Annunciation*, by Domenico Veneziano (*fl.* 1438, d. 1461), dating from the 1440s, is convincingly Albertian in certain respects (see Fig. 2.6). Not only do the orthogonals converge to a point in the distant door, but one can easily see a series of transversals. The large main panel of the altarpiece to which this small *Annunciation* belongs shows a Virgin and

**Fig. 2.6** Domenico Veneziano, *The Annunciation*, tempera on panel, 27.3 × 54 cm, *c.* 1445–1447, predella panel from an altarpiece for the church of Santa Lucia dei Magnoli (Florence), Fitzwilliam Museum, Cambridge.

Child enthroned against a rather complicated architectural background and accompanied by saints standing on a very complicated *pavimento*, seen at a very steep angle (see Fig. 2.7). However, Domenico Veneziano has not only made use of 'artificial perspective' (*perspectiva artificialis*) in giving a sense of space in his picture. He has also paid very careful attention to other elements in natural vision: we see the direction of the sunlight (falling from the upper right) and how the light is caught by the draperies. This more old-fashioned form of optics is, of course, also mathematical in its basis.

## Other perspective constructions

It is clear from later literary evidence, mainly dating from the sixteenth century, that the Albertian construction was not the only one current in the fifteenth. The chief alternative was a construction now known as the 'distance point construction'. Its simplest form is shown in Fig. 2.8. The problem is again to construct the perspective image of a square-tiled *pavimento* with one edge

**Fig. 2.7**  Domenico Veneziano, *Madonna and Child enthroned between Saints Francis and John the Baptist and Saints Zenobius and Lucy*, central panel from an altarpiece for the church of Santa Lucia dei Magnoli (Florence), 209 × 216 cm, *c*. 1445–1447, Galleria degli Uffizi, Florence.

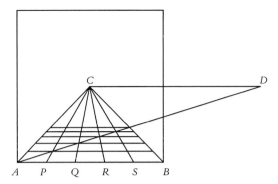

**Fig. 2.8**  Basic distance point construction.

along the ground line of the picture. We are given the centric point $C$. The line $CD$ is constructed parallel to the ground line $AB$. The length $CD$ is the distance of the eye from the picture. The point $D$ is called the 'distance point'.

The first stage of the distance point construction is as for the Albertian method, that is to put in the orthogonals by joining the division points on $AB$ to the centric point, $C$. Then $D$ is joined to $A$ and through the point in which $DA$ cuts $CB$ there is drawn a line parallel to $AB$. This line is the image of the furthest transversal, that is the line that marks the back edge of the pavement. Similarly, the intersection of $DA$ with each remaining orthogonal marks the level for each remaining transversal.

This construction can easily be seen to be equivalent to the Albertian one. We only need to imagine a slightly different form of superposition, in which the vertical through $M$, in the vertical section shown in Fig. 2.5(b), is positioned so that $M$ lies at the midpoint of $AB$ (and the vertical line passes through $C$). It must be pointed out, however, that this proof would almost certainly not have seemed satisfactory to a fifteenth-century mathematician. We shall see, in Chapter 5, that Piero della Francesca's proof of the correctness of his version of the Albertian method is very different in style from the arguments just given. Superposition was, it seems, a technique for getting things done, not an intellectual tool for proving mathematical truths. At least, that seems the likeliest explanation for the fact that it was not recognised that the Albertian method and the distance point method were mathematically equivalent. Indeed, as late as the 1620s, a Florentine called Pietro Accolti (*fl.* 1621–1625) misunderstood part of a serious mathematical discussion of perspective by Giovanni Battista Benedetti (on which see Chapter 7) and believed that it had been proved that of the two constructions only Alberti's was correct. Accolti accordingly gave the Albertian method the name '*costruzione legittima*', that is 'legitimate construction', which has unfortunately tended to stick to it. The perspective treatise in which Accolti coined this silly name is, perhaps appropriately, called *Deceiving the eyes* (*Lo inganno de gl'occhi*, Florence, 1625). Despite the racy title, the book is actually rather dull—this Pietro Accolti is no rival to his exact namesake better known as Pietro Aretino (1492–1556).

The version shown in Fig. 2.8 is what one might call an economical version of the distance point construction. A more extravagant method is sometimes recommended, and was probably used. It is shown in Fig. 2.9. In this variant, not only $A$ but all the points of division $P$, $Q$, $R$, $S$ are joined to $D$. For the highest transversal we again only obtain one point, namely the point of inter-

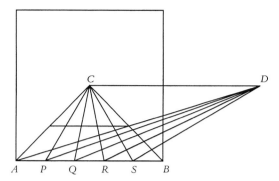

**Fig. 2.9**   Extravagant version of distance point construction.

section of *DA* and *CB*, through which we need to draw the transversal as a line parallel to *AB*. However, for the next transversal down we obtain two points, namely the points of intersection of *DA* and *CS* and of *DP* and *CB*, so the transversal is drawn more easily than the previous one. For the transversals lower still, we obtain still more points, for instance for the next one we have the points of intersection of *DA* with *CR*, of *DP* with *CS*, and of *DQ* with *CB*. Drawing the lower transversals is thus in principle rather easier, though the method is, of course, liable to show up the draughtsman's inadequacies by presenting a series of points that are visibly not collinear.

There is also a super-extravagant version of the distance point construction, shown in Fig. 2.10. Here every one of the transversals is given by more than one point, so each can be drawn simply by joining the appropriate points The redundancy that appears for the lower transversals will, as before, allow the draughtsman to check on his accuracy.

On a piece of paper, drawing lines parallel to the ground line is an easy task, and no doubt was equally easy in the fifteenth century. It is a more difficult job if one is working on a wall, as painters sometimes had to, for instance in preparing to paint a fresco. For one painter at least, the super-extravagant version of the distance point construction (shown in Fig. 2.10) did seem worth using: just such a construction was found on the underlying rough plaster when Paolo Uccello's

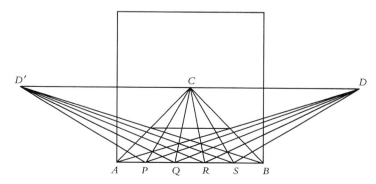

**Fig. 2.10**   Super-extravagant version of the distance point construction.

painting of the Nativity was taken off a wall in the church of San Martino alla Scala in Florence. Unfortunately, the picture is in a bad state—it was being removed from the wall because it was in need of attention from conservators—and the underlying drawing is impossibly difficult to show in a photograph. As was usual with drawings on the rough plaster (*ariccio*), the drawing was done in a red clay called *sinoper*.

Because they are so often drawn with this pigment, underdrawings are sometimes called *sinopie*. The red clay could be used directly, but to draw straight lines the usual practice was to rub a piece of string with a sufficiently sloppy preparation of the clay and water and then, holding the string taut with its ends in the appropriate places, twang it against the wall. In drawing on the finer-textured top layer of plaster (*intonaco*), lines were sometimes drawn by the same method but using a clean string, which merely left an indentation (this is called a 'snapped line'). *Sinopia* lines were easier to see, though, and artisans were not expected to mind about getting their hands dirty.

On a panel painting, the usual method of drawing was to use charcoal, which would show up clearly on the white *gesso* (essentially chalk) surface that was laid onto the wood to prepare it for painting. The charcoal could be wiped or blown off or brushed off with a feather if the draughts-man changed his mind. Once the design was decided upon, it was usual to ink it in and then remove the remaining charcoal. Consequently, apprentice painters also learned how to prepare charcoal and how to make ink. Sometimes, however, lines were actually engraved or scribed into the *gesso* (as was also done for drawing on paper).

In the painting by Giovanni Bellini (*c.* 1431–1516) shown in Fig. 2.11 there are engraved lines along all the edges of the tiles of the pavement. This looks like a splendidly clear example for the probable use of the distance point construction, but we must not jump to the conclusion that the actual construction process was carried out on the panel itself. As can just be seen in Fig. 2.11, the very edges of the panel show bare wood. The *gesso* surface has been broken away—because the original framing has been removed. Frames were fixed to the panel before the layer of *gesso* was applied. They were sometimes very elaborate, and it was not unknown for the craftsman who made and gilded the frame to be paid more than the craftsman (i.e. painter) who did the picture inside it. The *gesso* surface over the frame was necessary to make the wood smooth enough to receive the gold leaf, which was held in position with glue. Anyway, the result was that the painter could draw on his panel, but could not make lines going beyond its edges without encountering the frame.

Since Bellini's *Blood of the Redeemer* is a small picture, it was obviously reasonable to do some geometrical investigations using a photocopy of a photograph. Extending all the tile edges to the right, it was found that almost all of them converged fairly accurately to a definite point, at a level about the same as that of the centric point (which seems to lie at the point where the apparently solid halo touches Christ's head—and so much for Alberti's saying the centric point should be at the eye level of standing figures in the picture). The same sort of convergence point was found by extending the tile edges to the left. So we seem to have distance points. The viewing distance comes out at about two and a half times the width of the panel.

The procedure does not, of course, prove that the distance point construction was actually used in designing the pavement. One can carry out the same reconstruction procedure to find the position of the camera from a photograph. However, finding that two distance points can be constructed from the pavement does suggest rather strongly that some kind of correct perspective construction was used in the original design. Unfortunately, there has been some repainting of the pavement, which may have affected my estimate of the accuracy of the original construction. In

**Fig. 2.11** Giovanni Bellini, *Blood of the Redeemer*, oil on panel, 47 × 34 cm, about 1460–5, National Gallery,

any case, some inaccuracies would presumably have been introduced by the procedure of transferring a preliminary drawing to the panel.

There were a number of methods of transferring drawings to the surface prepared for painting. One method was to position the drawing on the surface and then run a suitable implement along the lines drawn on the paper, leaving indented lines in the surface. This, of course, also left marks on the paper drawing in the process. Another method, for which there is plenty of surviving evidence, was to cover the reverse of the drawing—partly or completely—with chalk or charcoal, and then, turning the paper back, transfer the drawing by running over the front of it again (as one does when using carbon paper). This method was used for copying drawings, and probably also in preparing small panels. A third method, well known from written sources, was to prick series of small holes along the lines in the drawing, lay the drawing on the prepared surface and pat along the lines with a pad filled with finely ground charcoal, thus leaving a line of charcoal-dust dots which could be joined up later if necessary. This method is known as *spolvero*, from the Italian word for 'dust'. It does not actually destroy the drawing used in the transfer process, but the patting with charcoal dust presumably left it looking rather unattractive. In fact, although a number of pricked drawings have survived, surprisingly few of them show signs of the use of *spolvero*. Drawings dirtied by *spolvero* were presumably more likely to have been thrown away, but it is also probable that designs were in fact pricked through onto another sheet of paper which was then used for the actual process of transfer. A two-step process would, of course, be inherently less accurate than a single-step one, so extra care would have to be taken. In any case, *spolvero* marks can be seen in many fifteenth-century frescos, though they only become visible when the paintings are viewed from very close to (or, of course, in suitably large-scale photographs).

Given its simplicity, the design for the *pavimento* in the *Blood of the Redeemer* could have been transferred by means of a very small number of *spolvero* dots—or, indeed, simply by pricking through a few points in the drawing to make indentations on the *gesso* suface. We accordingly have no chance at all of being able to decide which method Bellini used, and he may, of course, have used some quite different method that he happened to prefer. In any case, we have no way of knowing how he actually constructed the perspective of the pavement.

Sometimes we can do better than this. Our evidence for the quite widespread preference for the distance point construction does not rely on the small number of unfinished paintings, or infra-red photographs of finished works that show up appropriate underdrawing, or on the very small number of *sinopie* that give indications of how the perspective scheme was constructed. Instead, the evidence is mathematical: painters seem to show a preference for very short viewing distances. This may be partly explained by the fact that in practice, as artists must have noticed, the ideal viewing distance, built into the picture by the perspective construction, effectively acts as a minimum viewing distance. Pictures look suitably three-dimensional when viewed from too far away but make one feel uncomfortable if one stands too close. Trusting this effect is the most reliable method I have yet found for guessing at the ideal viewing distances of perspective pictures.

In their original settings, it would probably have been impossible for people other than the priest or privileged visitors to get very close to an altarpiece, so the artist could be sure that his minimum distance would be respected. Transgression is much easier in an art gallery, and it shows that many viewing distances are far smaller than would seem to be prescribed by the fencing-off arrangements for an altar or a chapel. An unexpectedly large number of pictures seem to have a viewing distance about equal to half the width of the panel.

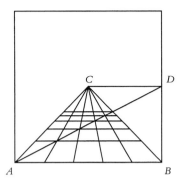

**Fig. 2.12** Distance point construction with viewing distance half the panel width.

The diagram for the distance point construction for this viewing distance is shown in Fig. 2.12. As can be seen from the figure, in this case, or indeed for an even shorter viewing distance, the construction can actually be carried out on the surface of the panel. At least, it can if one uses the distance point method. As can be seen from Fig. 2.4, the Albertian construction always requires the use of a point lying beyond the edge of the picture, since it is the distance of this point from the edge that establishes the viewing distance. Thus the marked preference for such short viewing distances is evidence for the use of the distance point method. It was clearly regarded as convenient to be able to carry out the construction directly on the panel. In the case of the fresco by Uccello which we have mentioned as using the super-extravagant version of the distance point construction, the two distance points do indeed lie on the outer rim of the picture.

Though we do not always know how he constructed them, the perspective schemes of Paolo Uccello's pictures are often decidedly visible. For instance, in the small panel of *St George and the dragon*, shown in Fig. 2.13, the plants, though clearly in a wild place, seem to have a wish to form themselves into a geometrical design that sketches a *pavimento*. In the large panel of the *Rout of San Romano*, shown in Fig. 2.14, the fallen soldiers and their broken lances seem to have contrived to make a square grid on the ground.

Uccello (1397–1475) is in fact known to have enjoyed playing with perspective effects. Giorgio Vasari (1511–1574), in his *Life* of Uccello tells the story that once when his wife called him to come to bed Paolo murmured 'Oh what a sweet thing this perspective is'. However, his nickname—'Uccello' means 'bird'—came from his habit of keeping so many birds of different kinds about his house. He loved to draw them. This concern with naturalistic detail is at least as visible as his concern with perspective. It may be that he thought of correct perspective as an aid to greater naturalism. Although it must be admitted that in many of his pictures the perspective does not actually look very naturalistic to a twentieth-century eye, accustomed as we all are to the instant correct perspective of photographs, it may have appeared otherwise to Uccello himself and his contemporaries.

## Alberti's check: an earlier rule?

We do not know the date or origin of the distance point construction. A hint may, however, be provided by the fact that after describing his own construction Alberti proposes that, to check that

**Fig. 2.13**   Paolo Uccello, *St George and the dragon*, tempera on canvas, 56.5 × 74.3 cm, National Gallery, London.

**Fig. 2.14**   Paolo Uccello, *Rout of San Romano*, tempera on panel, 181.6 × 320 cm, National Gallery, London.

all is in fact correct, one can draw the diagonal of the pavement, which should also be the diagonal of all the tiles lying along it. Now, this property of the diagonal is part of the basis of the distance point construction in the simple form shown in Fig. 2.8 and, perhaps less obviously, also of the more extravagant forms shown in Figs 2.9 and 2.10. However, if only the diagonal is drawn then one does not actually obtain a distance point. (We should remember here that for Alberti and his contemporaries, as for Euclid and his, the words 'the line $AB$' mean the line from the point $A$ to the point $B$, that is what would in today's terms be called a line segment. Thus the diagonal lies entirely within the trapezium unless it is prescribed as extended.)

Before proposing this check, Alberti has mentioned some earlier ways of constructing transversals. The check is presented as a way of proving they do not work (which is indeed the case). It thus seems rather unlikely that he actually knew any form of the distance point construction. On the other hand, he clearly does know about using a diagonal as a check, and the very lack of connection with a particular construction suggests something traditional, a rule of thumb the learned Alberti mentioned because it was known to artisans, and because someone had assured him it was actually correct.

In fact, the mathematics involved is so simple that it may look like commonsense to a mathematically-minded reader in the twentieth century. The historical significance of the use of the diagonal as a check lies in the fact that correct perspective did not end with Antiquity and did not recommence only after Brunelleschi made his discovery (whatever this may have been) in or shortly before 1413. There are a few fourteenth-century examples of perspective that appear to be totally correct. Figure 2.15 shows an example by Giotto. There seems to be no way of checking its correctness, but the effect of looking through an archway into a vaulted chamber behind it is

**Fig. 2.15**   Giotto, a view through into a chapel, fresco, chancel arch, Arena chapel, Padua, showing illusionistic rendering of architecture.

37

**Fig. 2.16** Ambrogio Lorenzetti, *Presentation of Christ in the Temple*, tempera on panel, dated 1342, Galleria degli Uffizi, Florence.

extremely convincing. A later example, by Ambrogio Lorenzetti (*fl.* 1319–1348), is however entirely open to mathematical analysis since it has an elaborate *pavimento*. The picture in question is shown in Fig. 2.16.

The small size of the reproduction impedes one's reading of the picture, since putting one's eye at the ideal viewing distance for the *pavimento*, namely about half the width of the panel, is not really practical unless one suffers from an appropriate degree of myopia. All the same, it is fairly clear from Fig. 2.16, and abundantly clear from the original painting, that the picture as a whole is not conceived in completely naturalistic terms. The single obviously illusionistic part is the *pavimento*. There seem to be seven tiles across the picture width, and there are four rows of tiles taking us back as far as the front of the altar. Once a *pavimento* is constructed, there is no dificulty in extending its width, so we shall look at the problem of making a four by four array of tiles.

The starting position is shown in Fig. 2.17(a). That is, the painter is imagined as having decided which area of the picture is to be covered by the pavement. For the sake of generality, and because Ambrogio Lorenzetti has an even number of rows but has chosen to put the centre of a tile on the centre line of the picture, we have made this area asymmetrical in the same way as the most visible bit of the *pavimento* in the *Presentation*. First of all, it is obvious that the back line of the *pavimento* will, like the front, need to be divided into four equal parts, one to represent each tile. Once these points of division are found, the orthogonal edges can be put in, as in Fig. 2.17(b). The next stage is to put in the diagonal line, shown as a broken line in Fig. 2.17(c) since it is merely a construction line not part of the final design. Now, it is obvious that in reality the diagonal cuts the orthogonals in the same points as the transversals, so this will be true in the picture also, and the transversals can accordingly be put in position by drawing lines parallel to the upper and lower edges of the trapezium, as shown in Fig. 2.17(d). Extending the width of the *pavimento* is very easy since the positions of the transversals are known and the orthogonals can be found by measuring additional units of the tile edges along the extensions of the top and bottom lines.

The steps of this proposed construction of the *pavimento* are very like those of the distance point construction, except that they do not involve either a centric point or a distance point. The construction does, however, involve one result which looks more like a theorem than an obvious

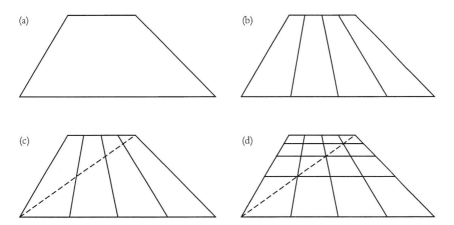

**Fig. 2.17** Drawing a 4 × 4 *pavimento* using a diagonal (note that the initial trapezium does not need to be symmetrical): (a) starting position; (b) orthogonals, found by making equal divisions along front and back lines; (c) adding the diagonal of the pavimento; and (d) putting in the transversals.

piece of commonsense knowledge, namely the assertion that the back (upper) line will need to be divided equally. Close inspection shows that the theorem one requires is not actually proved in Euclid's *Optics*. As far as I know, it is first proved in Piero della Francesca's treatise on perspective, which was probably written in the 1460s (see below); and since the result proves to be of some importance I propose to christen it 'Piero's Theorem'. To anyone except a mathematician with a professional aversion to claims of obviousness, the remainder of the stages of my postulated diagonal-line construction can surely be taken not to require any justification.

Artists' knowledge of a construction method something like this one, a knowledge passed on merely as an unexplained rule for making a convincing pavement, could account for the fact that pavements that seem to be mathematically correct are to be found in fourteenth-century pictures. Unfortunately, however, we have no literary evidence whatsoever concerning this matter (workshop rules did not necessarily get written down, of course). The story I have told of how such a rule might work is what historians of science rather warily call a 'rational reconstruction'. Mathematics is a highly rational subject, but we all know from our own experience that one can fail to spot things that, once spotted, seem obvious. The history of mathematics shows that on occasion even very talented mathematicians did not arrive at results which, using today's conventions and notations, seem entirely straightforward. In the case of pre-Brunelleschian *pavimenti* we are, moreover, dealing not with heavyweight mathematicians, but with craftsmen happy to follow established rules of thumb. So the knowledge of this particular 'rule' is a possible explanation for Alberti's reference to using the diagonal line and for the appearance of apparently correct *pavimenti* in a few fourteenth-century pictures. The 'rule'—if it existed—might perhaps have been passed down from master to apprentice since ancient times. The Ancients do indeed seem to have known some method of drawing in correct perspective, but we have (as yet) no very clear idea what it may have been.

This thought brings us back to Brunelleschi. Whatever he invented must presumably have seemed to him to be different from the rules of thumb employed in painters' workshops. We may perhaps doubt whether Alberti was entirely well informed about workshop practice, but Brunelleschi was himself a craftsman and must surely have had a good idea what rules painters actually used.

## Brunelleschi's invention

As we have already mentioned, works of art from the late 1430s onwards provide many examples of *pavimenti*. These may have been designed according to Alberti's method or by some method based on the diagonal that he proposed as a check. So our best evidence for Brunelleschi's perspective method must be in works made before Alberti's treatise appeared in 1435.

The earliest works to show orthogonals that converge to a point in a more or less convincing manner are not paintings, but relief panels by a friend of Brunelleschi's, the sculptor Donatello (1386–1466), properly Donato dei Bardi (he was a distant relation of the famous banking family, who paid for his education). One such panel, from the base of a statue of St George, is shown in Fig. 2.18.

It is not quite clear what one should try to measure on a relief. Moreover, Donatello's later relief sculpture, which often employs perspective to great dramatic effect, shows little concern for mathematical correctness, though there is sometimes enough to indicate that Donatello knew how to get the mathematics right if he felt it was appropriate to do so. So, all in all, it seems ill-advised to turn to Donatello's reliefs as evidence, though in the St George relief shown in Fig. 2.18 it is clear that

**Fig. 2.18** Donatello, *St George and the dragon*, relief, white marble, 39 × 120 cm, from the base of Donatello's statue of St George, Orsanmichele, Florence.

the orthogonal lines of the building on the right, including the faintly indicated orthogonal lines in the tiled pavement inside it, do converge to a point in St George's head. The convergence is too good to be reasonably explained as accidental.

Another, younger, friend of Brunelleschi's who was also an early user of perspective was the painter Masaccio (1401–*c.* 1428), properly Tommaso di Giovanni. Whereas Donatello's nickname is merely a diminutive, Masaccio's seems to be derogatory, and could be translated as 'big clumsy Tom'. The nickname is a little disconcerting, since Masaccio's art is far from clumsy. The style and strength of the figures in Masaccio's paintings are reminiscent of sculptures by Donatello, to whose work Masaccio may indeed have been reacting. This sculptural quality is clear even from small photographs, as can be seen in Fig. 2.19 which shows the fresco of the

**Fig. 2.19** Masaccio, *The Tribute Money*, fresco, Brancacci Chapel, Santa Maria del Carmine, Florence.

41

*Tribute Money* (whose figures are in reality about life size). This picture is part of a cycle of frescos, mainly relating to the life of St Peter, in the Brancacci Chapel in the church of Santa Maria del Carmine, Florence.

The *Tribute money* was probably painted in about 1426 or 1427. It is notable not only for the apparent solidity of its figures but also for the fact that we see them in space as real as they are. On the left, the landscape sweeps away to the distant hills, which seem to be a portrait of the hills behind Masaccio's native town of San Giovanni Valdarno. On the right, a more formal sense of space is established by the naturalistic perspective of the building. Assuming that the lines that look like orthogonals are intended as orthogonals, one may extend them to their point of convergence, which lies in the face of Christ. This looks very good on a small photograph. The restorers who worked on the picture in the late 1980s tried this investigation using the actual picture, thus avoiding the problem that a fine line on a small photograph would correspond to a strip several centimetres wide in reality. What the restorers found was that there appeared to be two points of convergence: one on the bridge of Christ's nose and one in the middle of His forehead. Since the figure is life size the error, if it was one, is considerable. In any case, Masaccio certainly achieves an effect of three-dimensionality, and it is clear that he must have employed a certain amount of mathematics in doing so.

Several of the other frescos in the cycle to which the *Tribute Money* belongs also give indications that orthogonals have been constructed rather exactly, probably by means of string attached to a nail hammered into the wall at the 'centric point'. However, the best example of Masaccio's use of perspective construction—that is the most obviously mathematical—is in the fresco of the *Trinity*, in the church of Santa Maria Novella (Florence). It is shown in Fig. 3.1 on p. 44.

The *Trinity* is highly praised by Giorgio Vasari in his *Life* of Masaccio. It was, in fact, Vasari's admiration that preserved the picture, for when Vasari, in his capacity as architect to the Medici, was ordered to remodel the interior of Santa Maria Novella, in response to the dictates of the Council of Trent, he arranged that the area of wall on which the *Trinity* was painted should be covered by a wooden altar. This, of course, hid the picture from view, but Vasari's alternative would have been to destroy it, so we must be grateful for his choice.

Since the *Trinity* fresco provides our best opportunity of studying Masaccio's mathematics, and hence making some better-educated guesses about Brunelleschi's, the next chapter will largely be devoted to an investigation of its perspective scheme. This investigation is based on measurements of the picture which two colleagues and I made, using scaffolding, in 1986 and 1987.

# Chapter 3

## THROUGH THE WALL: MASACCIO'S *TRINITY* FRESCO (*C*. 1426)

Masaccio's fresco of the *Trinity* is shown in Fig. 3.1. In his *Life* of Masaccio, in the version printed in 1568, Giorgio Vasari describes the picture in the following terms:

But the most beautiful thing in it, apart from the figures, is a semicircular barrel vault drawn in perspective, and divided into squares containing rosettes that diminish in size and are foreshortened so well that it seems as though the wall is pierced.

More recent descriptions have little to add to this, except to point out that the rosettes, no doubt painted onto the fresco surface after it had dried (using the technique called *a secco*), have now vanished. Their absence leaves the vault with a rather stark mathematical look that is more congenial to the taste of the twentieth century than it would have been to the much less austere taste of the fifteenth. From close to, one can see that a tangle of lines has been scratched into the surface of one of the coffers (using a brush handle?), presumably to indicate the placing of the now missing rosette.

The *Trinity* is a large picture—its human figures are life size—so only very approximate investigations of the perspective scheme could be made from normal-sized photographs. Thus, when a colleague started to ask me questions about the perspective, as we were standing in front of the picture one day in the summer of 1985, my response concluded '… but of course we should need scaffolding for that', and I assumed that the matter would end there. It turned out, however, that the enquiries originated from a friend of his whose job it was to look after the fabric of the church, and who could accordingly, with permission from his superiors, arrange for us to have the scaffolding we required. Permission was given, and we duly found ourselves confronted at close quarters with Masaccio's magnificent calligraphic brushwork and the mass of indentations and scribed, snapped and *sinopia* lines that he had used to mark parts of his design onto the fresh plaster. There was no way we could see the design as a whole, but we could take measurements, and do so with an accuracy comparable to that available to Masaccio. Thus it took no more than a glance to see the absurdity of the suggestion made by some art historians that God the Father's displacement from the central axis might be accidental. A *sinopia* line marked the axis, in the centre rib of the vault, and another *sinopia* line ran down the Father's nose. Their horizontal separation was instantly visible, and turned out to be 2.6 cm. Observations like this were rather good for our self-confidence—which turned out, as our work progressed, to need all the help it could get.

The mathematical means that we used in our investigation were, first of all, intended to discover mathematical properties of the picture. Until we knew something about that aspect of the painting there was no possibility of finding out anything of substance about the design procedures

**Fig. 3.1** Masaccio, *Trinity*, fresco, c. 1426. Santa Maria Novella, Florence.

Masaccio—and any possible helpers—had used in constructing the perspective. It turned out, in fact, that the array of modern tools we needed for our initial analysis was not particularly flashy. All the same, we were aware that such things as the cosine formula could have nothing to do with Masaccio's planning process in the 1420s. We set out first, in using such modern tools, to learn as much as we could about the results he had arrived at—conducting a kind of mathematical archaeology of the picture—and only later, and much more tentatively, did we try to consider what Masaccio himself had done to arrive at the results we had found.

## Looking for squares

We wanted to know the ideal viewing position for the *Trinity*, that is the position used in designing the perspective scheme. The three of us, and everyone else on whom we imposed the experiment of finding this point by moving forward and back in front of the picture, were in agreement that it was on the central axis, at a distance about equal to the width of the aisle of the church (at this place 686.25 cm). We were not sure about the ideal height for the eye, but there is no doubt that one is looking up at the 'chapel' in the picture, since one cannot see its floor, or the feet of the figures of the Virgin and St John. In principle, these facts were not news to anybody, but they seemed to be incompatible with the results of calculations based on measurements by two earlier scholars. One of these investigations had measured the actual picture (Kern in 1913), the other had made careful use of life-size photographs (Janson in 1967). These were the only known mathematical investigations of the perspective of the *Trinity*.

Once up against the wall, we used fine plastic-covered string to trace orthogonals downwards to the centric point. All the orthogonals had been marked on the plaster, in one way or another, before Masaccio began to paint the corresponding part of his picture. We found that most of the orthogonals converged rather accurately to a point on the central axis of the picture, as marked by Masaccio's *sinopia* line in the vault, and a little below the upper surface of the step on which the donors are shown as kneeling. (This point lies in a part of the picture that is the work of twentieth-century restorers—see below.) We found the height of the centric point above the floor to be about 172 cm, which is a reasonable eye height for a modern man. I later found it was suitable for a friend of mine whose height is about 188 cm. However, fifteenth-century people were generally smaller than that. Moreover, the present floor is nineteenth-century, and we do not know what the floor level was in the fifteenth century. All the same, it is clear that the *Trinity* has an element of illusionism not found in the *Tribute Money* (Fig. 2.19), whose centric point, being on the eye level of the standing figures in the picture, is about three metres above the eye level of a normal viewer. The height of the centric point in the *Trinity* is by comparison extremely naturalistic.

The most convenient pictorial element for finding the viewing distance of a picture is a suitably oriented square, that is a square with two edges parallel to the picture plane and two orthogonal to it. One can then carry out a kind of inverse version of the distance point construction, as shown in Fig. 3.2. (The reason why this diagram has been drawn upside down in relation to the diagram of the distance point construction in Fig. 2.8 will appear shortly.) The shaded trapezium represents the perspective image of a suitably oriented square. When produced, its two orthogonal edges will meet at the centric point $K$. The point in which a diagonal of the square intersects the horizontal line through $K$ is the distance point $D$. One could, of course, copy the pattern of the more extravagant version of the distance point construction (shown in Fig. 2.9) and use the second diagonal of

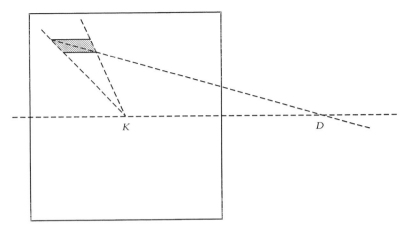

**Fig. 3.2** Inverse version of the distance point construction. The shaded trapezium is the perspective image of a square. Producing its edges gives the centric point *K* and finding where the diagonal meets the horizontal line through *K* gives the distance point *D*.

the trapezium to obtain another distance point on the other side of *K*. The length *KD* is the viewing distance of the picture.

The *Trinity* provides us with four squares oriented with edges parallel and orthogonal to the picture plane. These are the outlines of the flat slabs, confusingly called 'abaci', which lie on top of the four capitals of the columns at each corner of the painted chapel. In Ancient Roman architecture, such abaci seem usually to have had edges curving inwards, but Brunelleschi regularly uses square abaci in his architecture. The architecture shown in the *Trinity* has a Roman look to it, but is very much in the Brunelleschian manner, except for the use of a pretty pink where Brunelleschi uses the cool grey of *macigno*. Unfortunately, in the Brunelleschian manner, Masaccio's abaci are shown partly sunk into the wall—the correct architectural term for this is to say that such elements are 'engaged'. The abaci on the left are missing their left edges and those on the right are missing their right ones. However, it is clear that the abaci are in principle placed symmetrically on the columns, so one can make a reasonable estimate of the lengths of their upper edges.

Inspection of the fresco surface showed that the 'master' figures for the abaci were the upper left and the lower right. The other two had been drawn by taking over lengths with compasses (whose points had left indentations in the *intonaco*). The dimensions found for the upper left abacus are shown in Fig. 3.3, which is not to scale. Lengths are in centimetres.

The calculation evoked memories of secondary school trigonometry. The angle $\alpha$ was found from the right-angled triangle *PMK*, giving the value of $\angle QPS$, so that the cosine formula could then be used in triangle *PQS*, giving the side *QS*, allowing the sine rule to be used to find the angle $\theta$, which allowed the right-angled triangle *QLD* to be used to find *QL* and hence *KD* (= *ML*). The result was

$$KD = 553.11 \text{ cm (to 2 decimal places).}$$

The aisle width, that is the observed viewing distance, is 686.25 cm. So we had got a silly answer.

It is tolerably clear from Fig. 3.3 that this calculation of *KD* is unstable; that is, small differences in the length given to *PQ* will result in large differences in the value found for *KD*. Calculations

46

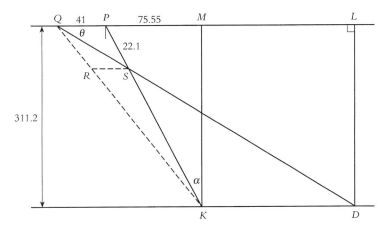

**Fig. 3.3** Masaccio, *Trinity* fresco, diagram of upper left abacus. Not to scale. Lengths in cm.

confirmed this was indeed so. However, the picture itself put limitations on how large one could reasonably make *PQ*, and its maximum value appeared to be 43.85 cm. This value looked too large when checked against the painting some months after the calculation had been done, but one can construct an argument for accepting it. Taking *PQ* = 43.85 cm gave us *KD* = 594.41 cm. Standing in front of the *Trinity* confirmed that this distance, which is about 90 cm less than the aisle width, is in fact too short. The excessively short distances found from the front left abacus were, however, reassuring in one respect: they agreed quite well with the excessively short distances found by our predecessors (Kern in 1913 and Janson in 1967). We were undoubtedly in intellectual trouble, but we were in thoroughly distinguished company.

The lower abaci, those at the back of the painted chapel, had not been used by either of our predecessors. The right one provided a very welcome surprise: its outline was shown completely in scribed lines and the orthogonals, which went as far as the edge of the *giornata*, were aligned very accurately towards the centric point. So we gratefully measured all four edges of the trapezium and prepared to do calculations using both diagonals. The measurements are shown in Figs 3.4 and 3.5, which are not to scale. Lengths are in centimetres. Like the previous calculation, these two are also unstable, but in these cases we are not having to guess any lengths, so the source of error is our error in measuring, which is certainly not more than ±0.5 mm. There is no possibility of the instability of the calculation being able to explain away the lengths found for $KD_l$ and $KD_r$, which were

$$KD_l = 360.61 \text{ cm}$$
$$\text{and } KD_r = 345.66 \text{ cm}.$$

These are ridiculous. The aisle width is 686.25 cm. In fact, these results are so ridiculous that I had seen them coming from my rough measurements of an A4 photograph, carried out by drawing the diagonals of the abaci on a photocopy.

To sum up: investigating the squares had added another element to the problem, and had provided a hint towards a solution only in the very limited sense of suggesting that Masaccio had not regarded the squares as primary elements in producing a perspective illusion. As Vasari had noted,

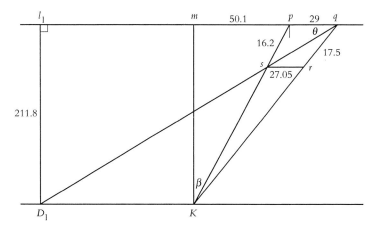

**Fig. 3.4**  Masaccio, *Trinity* fresco. Lower right abacus with dimensions, and distance point to the left.

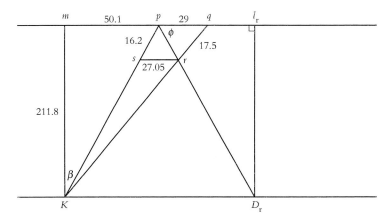

**Fig. 3.5**  Masaccio, *Trinity* fresco. Lower right abacus with dimensions, and distance point to the right.

the vault was what took the eye through the wall. No doubt Masaccio had reckoned on this being so. However, the squares being rendered so inaccurately did suggest that, whatever method Masaccio had used, it did not privilege the rendering of squares in the way we find in the Albertian and distance point constructions. There simply is no visible *pavimento* in the *Trinity* fresco.

Using squares to reconstruct the perspective scheme has the advantage of allowing us to ignore the question of the exact position of the picture plane in regard to the scene portrayed. The value found for *KD* will simply find the distance of the eye from the actual picture surface. As we have seen, this method worked quite well for Bellini's *Blood of the Redeemer* (Fig. 2.11), but gave completely unsatisfactory answers for the *Trinity*. In turning to a more basic method of investigating Masaccio's perspective, namely considering simple projection, as shown in Fig. 3.6 below, we needed to decide where the picture plane was imagined to be in relation to the architecture shown

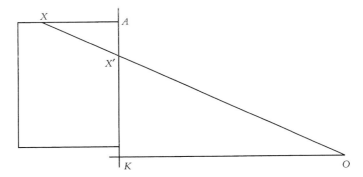

**Fig. 3.6**    Masaccio, *Trinity* fresco. Simple projection for elements of the barrel vault.

in Masaccio's composition. The answer may seem obvious from Fig. 3.1, or even from looking at the picture itself in its present state. However, we know that we have no longer got all of what Masaccio painted. Restorers have tidied up all four of the edges. In particular, almost everything that is now visible below the level of the donors' step (including the area containing the centric point) is a modern reconstruction, though enough original plaster remains to provide some justification for what has been shown.

From scaffolding, there is no difficulty whatsoever in identifying restored areas: the restorers follow the prescription of Renaissance manuals in getting their plaster surface flat, Masaccio almost never does so. Perhaps his preference for working on a surface that was not entirely flat is what earned him his apparently derogatory nickname? In any case, though there is no danger of mistaking the restored areas for original ones, that does not help a great deal in deciding just what we may have lost from Masaccio's original framing for his picture. Since art historians are not of one mind on the matter, we decided to take the apparently commonsense view that the picture plane is that of the front of the pink moulding round the arch and the front surfaces of the abaci on the front pair of columns. It will be noted that this means that the kneeling donors are on the spectator's side of the picture—which is not a common motif in the art of this period, though some comparable examples can be found.

## The vault: orthogonal section

Having decided on a position for the picture plane, we were able to investigate the perspective rendering of the vault by considering simple projection, as shown in Fig. 3.6, in which *A* represents the highest point of the outer edge of the semicircular moulding. In the raking light provided by our bright viewing lamp, it was easy to see the marks Masaccio had made on the fresh plaster to guide him in painting the vault. Figure 3.7 shows part of the left half of the vault, which was the first part to be painted. On the right half, there are fewer scribed and snapped lines and some lengths have been taken across from the left with compasses, whose points have left small indentations in the plaster.

We had plenty to measure. Measurements were all made directly from *A*, giving us values for *AX'*, working down the *sinopia* plumb line that Masaccio had provided to mark the axis of the

**Fig. 3.7**   Masaccio, *Trinity* fresco, left side of vault.

picture. As can be seen in Fig. 3.7, this line does not lie exactly in the middle of the central rib, but it is certainly the axis since the centric point lies on it. We measured everything there was. That is to say we measured the positions of the arcs marking the centres of the ribs and coffers as well as those marking their outlines in the painting. These arcs had clearly been struck with compasses— either beam compasses or an improvised version of them—and the arcs were not concentric. The positions of the centres all lay within the figure of God the Father, that is in an area which at the time would have been rough plaster (*ariccio*). Looking at how roughly Masaccio had treated his *intonaco*, we were happy to imagine him hammering nails into the *ariccio* and marking centres for re-use in the drawing of the right side of the vault, where the arcs for edges join up exactly but the centres of the ribs and coffers are not marked.

Measuring all these values of $AX'$, and finding $AK = 416.8$ cm, we were able to use the similar triangles $XAX'$, $OKX'$ to give us values for $AX$, in terms of the ideal viewing distance $KO$, which we called $d$. This time, the only problem was a tendency to run out of alphabet. We even managed to feel some sympathy for Piero della Francesca's habit of numbering rather than lettering points in perspective diagrams (see Chapter 5).

The distribution of ribs and coffers that we obtained is shown in Fig. 3.8, where the scale is in units of $0.01\,d$, and $A$ corresponds to the point $A$ in Fig. 3.6. From the point $E$ backwards the spacing of the alternating centres of ribs and coffers is very nearly even (the maximum 'error' being 4%). So we decided that for the purpose of our initial analysis we could usefully think in terms of a unit cell, consisting of one rib plus one coffer. Its dimension $u$ is given by

$$u = 5.76 \times 0.01\,d.$$

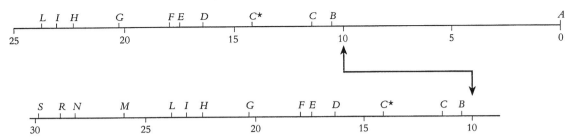

**Fig. 3.8**  Masaccio, *Trinity* fresco, section of vault with measurements deduced from calculation.

However, *E* is too far back in its rib and adding the length of a unit cell in front of it takes us almost exactly to *C*. So *C* should have been the centre of a rib whose marked centre is a little further forward, at *B*. A glance at the corresponding part of Masaccio's painting (Fig. 3.1) shows that what has happened is that the rib in question has been moved up a bit so as to be partly hidden by the pink moulding. Presumably Masaccio did not want a heavy pale grey curve to come so close to the pink one. Having moved this first rib, he then moved the next one up a little also, to stop the first coffer looking excessively large—but left us the scribed arc through what should have been the centre of this second rib (our *E*). The calculations for the radii of the arcs through points *C*, *C★*, *D*, *F* must have been done after Masaccio took his decision about moving the ribs. Which means that if anyone was helping Masaccio with his mathematics then the helper also knew about this departure from mathematical correctness. Of course, Masaccio may not have needed such a helper, but there is universal agreement among art historians that, if he did, the most likely candidate for the task is Filippo Brunelleschi, whose style is so visibly present in the architecture shown in the *Trinity*.

Since Brunelleschi is known to have been interested in tidy series of proportions, we found it rather satisfying that the results summarised in Fig. 3.8 gave us coffers and ribs with depths in the ratio 3:1. This ratio seemed plausible as a subsidiary element in the original design of the architecture of the imaginary chapel. However, what we were really after was a value for *d*. For this we had to look at elements parallel to the picture plane.

## The vault: transverse section—and a viewing distance

Diagrams in art history books quite often show how Masaccio could have divided the arc of his vault into eight equal parts to obtain the positions of the orthogonal ribs. In fact, the division is not equal. Even in a photograph as small as our Fig. 3.1 one can see that the lowest row of coffers on each side is larger than the higher ones. This is, of course, even clearer in the original fresco. From scaffolding one can hardly fail to notice that the coffers in the lowest row are almost exactly one and a half times as wide as the others. The difference does not obtrude itself upon the normal viewer, on the floor of the church, because the lowest row of coffers is in shadow, but it is none the less real in mathematical terms.

As with the re-spacing of the transverse ribs, the re-spacing of the orthogonal ones has an obvious explanation in pictorial terms. As painted, the lowest ribs run between the column capitals and the hands of Christ on the Cross. The architecture has, after all, been designed not for its own sake but as a setting for the figures. This seems entirely obvious to me, but I should perhaps warn my readers that telling it to an art historian, who had asked my opinion about a mathematical way in which the uneven divisions might have been obtained, made him emit a sound somewhat like that of crashing gears. He believed me, though.

The six remaining coffers, three on each side of the central rib, appear to be equal, and our measurements confirmed that this was indeed so. As is well known, there is no exact way of dividing an arbitrary arc into six equal parts. However, division of the circular edge of a wheel blank into any number of equal parts is a standard task for a clockmaker. It is carried out by guessing the correct compass opening, 'stepping' the compasses round the rim, adjusting the opening if necessary, 'stepping' again, and so on as many times as required. With practice, such tasks can be done very easily and quickly. We know that Brunelleschi made clocks. So there is no difficulty in finding a formal way in which Masaccio may have made his division into six—if the matter is seen to stand in need of such explanation. Possibly it does, since the division is in fact extremely accurate, no errors at all being detectable by our use of a scale marked in millimetres.

The diameter of the barrel vault can be measured directly, since we have assumed that the front surface of the pink semicircular moulding lies in the picture plane. Checking the number of little steps in the moulding shows that what is visible at the back of the vault has the same number as what we see at the front, so the outer edge of the moulding corresponds to the surface of the vault. The diameter of this semicircular edge is 211.55 cm. The unit of measurement employed in Florence in the fifteenth century, and for several centuries before and after it, was the *braccio*, literally 'arm' and similar to the English measure called an 'ell'. It was originally a measure for cloth. Since its international trade in cloth was a very important component of the Florentine economy, the length of the *braccio* was fixed by statute. It was equivalent to 58.36 cm. The diameter of Masaccio's vault is thus very close indeed to $3\frac{3}{8}$ *braccia*.

To decide on the size of the coffers we have to decide how they are meant to relate to the vault. Various possibilities were actually tried, but the one that could best be coaxed into making sense mathematically was that the circumference was intended to have eight equal coffers. Any number of art historians have apparently read the picture in this way, so assuming Masaccio meant them to do so has the added advantage of allowing us to be polite to all parties. If there are eight coffers, we still need to settle the number of ribs. The pink moulding at the top of the wall joining the lower edge of the vault stops us from seeing what is going on. My attempts to find a suitable piece of actual architecture by Brunelleschi to compare with this one were unsuccessful. Peering closely at the fresco suggested that half a rib was visible at the back of the vault.

If we assume that there is a half rib at the base of the vault on each side, then the total circumference contains eight ribs plus eight coffers, that is, eight of what we have called the unit cell, so the dimension of the unit cell is given by

$$u = 0.7118 \; braccia.$$

From our investigation of the orthogonal section we have $u = 5.78 \times 0.01 \, d$, so we obtain

$$d = 12.38 \; braccia,$$
$$= 721.3 \text{ cm}.$$

The aisle width, our observed viewing distance, is 686.25 cm, that is about $11\frac{3}{4}$ *braccia*, so this calculated viewing distance is too large.

If we assume there are complete ribs at the base of the vault at each side, then the circumference comprises $8\frac{1}{3}$ unit cells and we thus have

giving us
$u = 0.6833$ *braccia,*
$d = 11.87$ *braccia,*
$= 692.5$ cm.

This is a mere 6.25 cm larger than the aisle width.

In both calculations, we are assuming that the coffers are square, in the sense that they would come out square if the cylinder of the barrel vault were unrolled to become a plane surface. Given what is known about Renaissance architecture, this seems the most reasonable assumption to make, though there is an alternative, namely to take the ground plan of the painted chapel to be square and allow the coffers to be rectangular. There is absolutely nothing in our mathematical analysis that provides any grounds for choosing between these alternatives—but our choice was in fact for square coffers and it is left as an exercise for the reader to explore the mathematical consequences of the alternative choice.

Measurements of rib and coffer widths had confirmed the 1:3 ratio in their sizes found from our investigation of the orthogonal section, so the values found for $u$ allowed us to calculate the sizes of individual components. The smaller value of $u$, corresponding to the best fit for $d$, is $u = 0.6833$ *br*. This gives a coffer width of $0.5127$ *br*, that is about half a *braccio*. We can get even closer to half a *braccio* if we actually put $d$ equal to the aisle width, 686.25 cm. This then gives us a coffer of $0.5072$ *br*.

## The dimensions of the vaulted area

The complete length of the vault can also be found by simple projection, as shown in Fig. 3.9.

Figure 3.9 shows the method applied to the height of the vault. It can also be applied to the width of the vault and to the heights of the top surfaces of the abaci on the columns at back and front. The three values of $q/d$ that we obtained from such measurements were 0.4780, 0.4650 and 0.4693. As the reader can easily check, the values corresponding to $d = 692.5$ cm and those

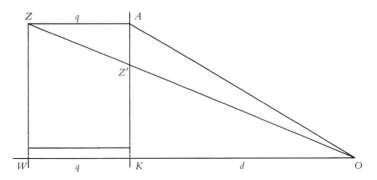

**Fig. 3.9** Masaccio, *Trinity* fresco, orthogonal section to show calculation of total length of vaulted area.

corresponding to $d = 686.25$ cm, combined with the corresponding possible values for $u$, all end up giving us a vaulted area with a ratio of depth to width of about 4:3. Historians of architecture may find this less 'Brunelleschian' than a square, but it is at least related in a simple mathematical way to the ratio between coffer and rib widths, which is 3:1. In any case, it seems reasonable to suppose that it was the proportions of the overall shape of the architectural structure that were the basis of the design. Unfortunately, what the picture shows does not permit us to make many measurements that are direct indicators of the overall shape. We are accordingly on surer ground for the smaller, repeated, elements. This is another example of surviving evidence tending to provide answers to one's less interesting questions. Sod's Law is operative in history as in science.

The story so far makes it clear (I hope) that Masaccio's various deliberate adjustments in portraying the architectural setting, together with the choice that has to be made about the shape of the coffers, make it mathematically impossible to give an authoritative plan and section of the architecture. There is, however, a further possible check for the reasonableness of our proposed reconstruction. We may ask whether it could accommodate the figures shown in it in the painting.

## The figures

We cannot see the feet of the Virgin or of St John, so these figures will not help us. This leaves the three persons of the Trinity. The two men and a dove do seem to have been shown as if they were normal-sized for the forms in which they are represented. Our best guide would be Christ, since His true height is known, in the sense that the height of a perfect man was taken, by Alberti among others, to be 3 *braccia* (175.08 cm). Christ was, of course, perfect in height as well as everything else. If we make the theologically more dubious assumption that like Father like Son, we may guess that Masaccio also showed God the Father as 3 *braccia* in height. His height in the picture is 155 cm, and Fig. 3.10 indicates the nature of the ensuing calculation. In fact, as already mentioned, a vertical section through the central axis of the figure of the Father, as indicated by Masaccio's *sinopia* line on His nose, would not pass through $AK$ but 2.6 cm to the left of it. This small deviation has been ignored in our diagram, since it does not have a noticeable effect on the calculation, which depends on the properties of various pairs of similar triangles. We find

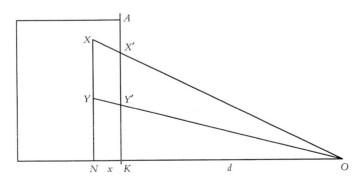

**Fig. 3.10**   Masaccio, *Trinity* fresco, orthogonal section showing God the Father in the chapel.

$$x = 0.1295 \, d.$$

Since the depth of the chapel is about 0.47 $d$, this means that the Father is standing between a quarter and a third of its length back. Taking the viewing distance $d$ as the aisle width, 686.25 cm, the distance between the front of the entrance arch and the figure of God the Father is 88.9 cm. This seems to leave adequate room for the remaining figures, particularly since St John's shadow is cast on the side of the structure supporting God the Father, thus indicating that he is standing quite close to it. The distance between God the Father and the back of the chapel comes out at about 233 cm, which would be sufficient to allow the structure on which He is standing to be identified as a tomb, namely the tomb of Adam. This identification is in accord with the most widely accepted interpretation of the picture, namely that it is of the iconographic type known as 'The Throne of Grace'. That element in his picture would almost certainly have been prescribed for Masaccio by his patrons, so it must be expected to have been at least fairly conventional.

## Explaining away the shapes of the abaci

The adjustments Masaccio made to the positions in which transverse and orthogonal ribs of the vault were shown are matters of demonstrable mathematical fact. Moreover, perhaps they should not surprise us very much: a coffered barrel vault is not a background one would recommend for a photograph. The best explanation I can volunteer for the incorrect shapes of the abaci is that they also have been subjected to adjustment for reasons to do with the composition of the picture as a whole, though the adjustments are much smaller than for the ribs.

Calculation shows that for the abaci on the front columns, seen from a viewing position at a distance of the aisle width, 686.25 cm, the length of the receding edge should have been 19.2 cm (assuming the maximum reasonable size for the edge, namely 43.85 cm). The receding edge as painted is very precisely directed towards the centric point, as it should be, but its length is 22.1 cm. The adjustment of a little under 2 cm is small, but the calculation for the viewing distance is, as we have seen, decidedly unstable, so the change shows up as a shortening of the calculated viewing distance by about 95 cm.

Why should Masaccio have lengthened the orthogonal edges of the front abaci? Well, these edges play an important part in his composition because they are close to, and therefore serve to emphasise, the two lowest orthogonal ribs (see Fig. 3.11). These are the ribs which link the column capitals to the hands of Christ, and their positions have been adjusted precisely to fulfil that purpose. As painted, the orthogonal receding edges of the front abaci are seen to end against a transverse rib, thus causing minimal disturbance to the viewer's reading of that part of the pattern of ribs in the vault. Had these edges been shorter, the abaci themselves would have been much more conspicuous, thereby distracting the eye from the pattern made by the ribs. That is, at any rate, the best I can at present suggest by way of explanation for the shape of the front abaci.

For the shape of the abaci at the back, the adjustment is much greater, and may have been made to achieve a satisfactory relation between the abacus edges and the volutes in the column capitals. This would seem to be a matter of minor importance, but maybe Masaccio thought the same about adjusting the edges of the abaci? In any case, numerical investigations do not suggest that the back abaci have been constructed as scale copies of the front ones, so that apparently obvious explanation is no help. In fact, when one checks sizes it becomes clear that the back capitals

**Fig. 3.11**  Masaccio, *Trinity* fresco, top left abacus to show edge against rib.

themselves are not scale copies of the front ones—as they should be—though the two back capitals are the same as one another. Almost unbelievably, since one might have expected such an 'error' to act strongly against the spatial illlusion, the vertical depths of the back abaci are apparently identical with those at the front. Is one to suggest Masaccio may have noticed the 'size constancy' effect by which, according to today's perceptual psychologists, we tend to see things as in some sense 'the same size' if we for some reason actually know that they are in reality the same size? It seems more likely that he was in fact thinking of repeated elements in the picture plane rather than repeated elements in space. Thus the relative sizes of the column capitals may have been dictated by his wish to have a series of three pink areas on each side of his picture, combining to sketch a surface V to balance the pattern of the converging orthogonals. This is known (to me) as the Pink Splotch Theory. It is decidedly nihilistic but I very much doubt whether a more respectable theory can be arrived at by the exercise of yet more twentieth-century mathematics.

## Masaccio's mathematics

What of Masaccio's own mathematics? It seems possible that our failure to find any mathematical pattern in the relationship between the back column capitals and the front ones is an indication, not that Masaccio made adjustments, but that he did not actually make any calculation at all. The scaling calculations that he should have done—if in search of optical exactitude—depend not on any new mathematics but upon the old-fashioned Euclidean optics that was well known to natural philosophers of the day. The one bit of optics to which he does seem to have paid careful attention is the convergence of the orthogonals. The lengths of all orthogonal edges of the abaci are wrong, but their alignment to the centric point is very accurate. The use of snapped lines for many of the orthogonals, including those in the moulding at the base of the vault, suggests that they were put in by means of a string attached to a nail driven into the wall at the centric point. As was mentioned in Chapter 2, there is evidence that Masaccio used a similar method in some of the pictures in the fresco cycle in the Brancacci Chapel.

The marks Masaccio left on the *intonaco* of the *Trinity* fresco also provide evidence for his use of other kinds of mathematical construction. For instance, there are tiny square-based pyramids on the inside surface of the moulding round the entrance arch. Each of them shows one face as an equilateral triangle, of side 29 mm, and the vertices of the triangle are marked with small indentations. All the triangles are exactly the same. Masaccio must surely have used triangular compasses (that is compasses with three points rather than two) or perhaps a template of some kind. Though the vertices are put in exactly, it is not quite accurate to describe the shapes as pyramids, for their outward-pointing edges are not quite straight. Masaccio has drawn each of them with a curved flick of dark pink paint, making shapes like the pinnacles familiar from contemporary metalwork. In the small area of modern plaster in this part of the picture, the restorers, exercising hindsight, have given a more correctly Roman effect by making neat straight-edged pyramids, scribing complete edges into the plaster for the purpose.

Another piece of rather more advanced craft mathematics is visible in the two pink roundels (*paterae*) beside the capitals of the pilasters near the top of the picture. The right-hand one is shown in Fig. 3.12.

The arcs marking the outer rims and the tips of the ridges of the flutes have been scribed with compasses, whose other point presumably made rather a mess in the centre since that area of plaster has been retouched. The circle through the tips of the flutes has then had their individual positions marked on it with small indentations. A series of smooth freehand sweeps of pink paint join these points to the centre, making the shapes of the actual flutes. There is a wonderful contrast between the mathematics and the art. The former is, however, also worth a second glance. The *patera* has eighteen flutes. Sixteen would have been much easier to construct, if one were starting merely from a circle. On the other hand, the division into eighteen is very easy if one has access to an astronomical observing instrument such as an astrolabe or a quadrant: the limbs of these instruments were divided into degrees, so division into eighteen equal parts simply meant working

**Fig. 3.12**  Masaccio, *Trinity* fresco, top right patera.

round 20° at a time. We know that from the seventeenth century onwards the makers of astronomical instruments regularly used a circular piece of apparatus called a 'division plate', which had marks on it for division into degrees and could be used to transfer the appropriate divisons to a blank. Clockmakers of this time also used division plates, marked with almost any number of equal divisions, for constructing the wheels of clocks and watches. It is not known whether division plates were used in the fifteenth century. Perhaps Masaccio's eighteen-lobed *patera* should be taken as evidence that they were. At the least, the use of eighteen divisions suggests some contact with the world of astronomical instruments or clocks. Since Masaccio himself is believed to have been trained as a painter, the likeliest mediator for such contact would seem to be Filippo Brunelleschi.

**Fig. 3.13** Masaccio, *Trinity* fresco. Lower ends of ribs joining moulding at base of vault, showing the hands of Christ and the Father.

This is not to say that Brunelleschi was actually responsible for any of the mathematical construction we find in the *Trinity*, or for whatever preliminary mathematical work was done before work on the actual picture began. For all we know—and what we know about Masaccio is very little, since he died so young—Masaccio may have done all his own preliminary constructions. We do know, however, that it is Brunelleschi whom everyone credits with having invented a 'rule' for mathematically correct perspective. So, unless we are to propose the implausible hypothesis that Masaccio had secretly invented a 'rule' of his own, it seems reasonable to hope that Masaccio's construction for the *Trinity* was as Brunelleschian as the style of its architecture.

We have already noted the absence of an explicit *pavimento* in the *Trinity*, but one could imagine a quasi-*pavimento*, or at least its transversals, formed by the horizontal diameters of the transverse ribs of the barrel vault. This kind of construction for the ribs is suggested in several books on Brunelleschi, but Fig. 3.13 shows how implausible the suggestion is. The scale of the photograph can be judged by the fact that the hands are life size. The brushwork is a reminder of how difficult it was to keep our minds on Masaccio's mathematics. As can be seen, the points where the ribs meet the horizontal pink moulding are ill defined and very close together. In fact, the compass sweeps that mark the edges of the ribs (only the edges, since we are on the right-hand side of the vault) do not extend as far as the snapped line marking the position of the moulding, nor are there any marks along it to indicate where they should meet it. It would clearly have been an inherently awkward process to use a construction working up from such closely spaced points, and there is no indication that Masaccio has done so. Moreover, if he had used an Albertian construction, or any other method that privileged the construction of squares, it is hard to believe that he could then have been so casual about the shapes he actually gave to the only four true squares in his composition. After all, he was very careful with the geometry of the tiny pyramids and the *paterae*. On the present evidence, my theory is that what Brunelleschi had discovered was the convergence of the orthogonals to a point which was the foot of the perpendicular from the eye to the picture plane. The rest of what Masaccio is using (or not) would seem to be a matter of ordinary Euclidean optics, such as the simple projection shown in our Figs 3.6, 3.9 and 3.10. I should also bet on the use of a certain amount of guesswork, getting it to look right and so on—in the style of the explanation that I called the Pink Splotch Theory. We are dealing with a painting not a theorem.

## Visually right and mathematically wrong

How did Masaccio know that he could get away with the departures from mathematical correctness that we have found in the *Trinity* fresco? There can be no doubt that he has got away with them, triumphantly. In case readers are wondering, I may as well say that the only effect the above investigation has had on my personal apprehension of the picture is that I still occasionally dream about Masaccio's brushwork (and sometimes see the pictures he would have painted had he lived on into the 1460s). As visual illusionism the *Trinity* undoubtedly works, and Masaccio, who was then a young man with a career to make, must surely have had good reason to suppose it would. As we saw in our last chapter, frescos had to be planned.

Vasari's description provides us with one possible hint by mentioning the figures. Their sculptural solidity insists that the eye construct a space around them. Masaccio could have recognised

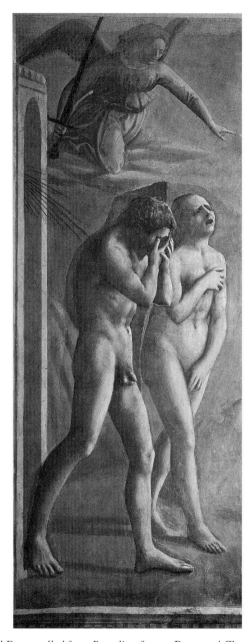

**Fig. 3.14** Masaccio, *Adam and Eve expelled from Paradise*, fresco, Brancacci Chapel, Santa Maria del Carmine, Florence. The figures are about life size.

this kind of effect by looking at Giotto's frescos in the Bardi and Peruzzi chapels in Santa Croce (about half an hour's walk from Santa Maria Novella).

Masaccio might also have discussed the matter with Donatello. Sculptors generally worked with their piece held at an angle, at least partly because the finished sculpture would become part of a building and would consequently, in most cases, always be seen at an angle, sometimes a very steep one. Donatello took good care to allow for this effect in his sculptures. That is, he was adept at allowing for natural optics in making things look right. As mentioned in the last chapter, his use of perspective is dramatically effective but in general very far from mathematically correct. In particular, when they are to be placed high up, Donatello's sculptures are often 'distorted' so as to make their lower part longer than it should be, this distortion then being compensated by the effect of their being viewed from below. Masaccio seems to have done something rather like this in painting the scene showing the expulsion of Adam and Eve from Paradise in the fresco cycle in the Brancacci Chapel. The picture is shown in Fig. 3.14. It is famous for its masterly portrayal of the naked human body—though the colour is now rather pink, probably due to the chemical changes brought about by the heat of the fire which consumed the rest of the church in the eighteenth century (the extent of the colour change has become apparent only recently, as a result of cleaning). The contrasting heads are particularly impressive. The legs, however, seem too long, and this may perhaps be explained as Masaccio's way of allowing for the fact that the picture is painted very high up on the wall, so that one can only see it from a reasonable distance by looking up at a sharp angle. The photograph shown in Fig. 3.14 was, of course, taken from scaffolding. From the floor of the chapel, one loses something of the force of the heads, but the legs look more or less natural. So I should like to suggest that perhaps the reason why Masaccio knew he could get away with the mathematical inexactitudes of the *Trinity* was because he had been thinking such things over in the company of Donatello.

Like other artists of the fifteenth century, Masaccio and Donatello were interested in a form of truth that was essentially visual rather than mathematical, though mathematics might be used in attaining to it. That a picture so impressively visually correct as the *Trinity* can turn out to be mathematically faulty is a warning against confusing artist with mathematician. In Masaccio's case, we can be sure he had access to good advice about the mathematics of perspective. For other artists we cannot be so sure. In any case, we know them only as users of mathematics, adding some new mathematical techniques to the traditional skills of their craft. It is accordingly very difficult to be sure how well they understood the rules they were using. There is, however, one fifteenth-century exception to this generalisation: Piero della Francesca, who is the subject of my next two chapters, was known in his lifetime as a competent mathematician as well as a painter. His work is thus of exceptional interest in exploring the relations of mathematics and art.

# Chapter 4

# PIERO DELLA FRANCESCA'S MATHEMATICS

Piero della Francesca (*c.* 1412–1492) is acknowledged as one of the most important painters of the fifteenth century. His pictures are notable for their characteristically Renaissance quality of seeming to portray real figures in real pictorial space (see Fig. 4.1). Piero is, however, unique in also having an independent reputation as a highly competent mathematician. As we have seen from our investigation of Masaccio's *Trinity*, it is extremely perilous to attempt to use works of art as evidence of an artist's mathematical skills. Luckily, for Piero the case is entirely different. We have three mathematical treatises by him—Vasari says he wrote 'many' such works—and these establish that as a mathematician he was indeed both competent and original. How far his mathematical skill can be seen in his art is a more difficult question, so it is with his mathematics that we shall begin.

Piero della Francesca was the son of a dealer in leather and other commodities (including the dyestuff indigo, which was important to the Florentine trade in cloth). His family was clearly quite prosperous and Piero, like his brothers, seems to have been given an education that would equip him to join the family business. That is, he probably attended an abacus school. In any case, Vasari tells us that Piero showed mathematical ability in his youth, and might have made mathematics his profession had he not decided to become a painter. As it was, Piero was at first apprenticed to a local painter, working in and around his native city of Borgo San Sepolcro (now called Sansepolcro). When he reached his early twenties, Piero shifted his sights to the nearby city of Florence, no doubt rightly recognising that was where the action was. As it happened, however, he does not seem to have lived in Florence very long. His work was mainly done in towns which were of lesser artistic importance—though the survival of Piero's pictures has now made some of them places of artistic pilgrimage.

We do not know a great deal about the circumstances under which Piero's three surviving mathematical treatises were written. Even their dates of composition are unknown, and we are not sure what titles Piero gave them. The titles by which they are now known are: *Abacus treatise* (*Trattato d'abaco*), *Short book on the five regular solids* (*Libellus de quinque corporibus regularibus*), and *On perspective for painting* (*De prospectiva pingendi*). The perspective treatise, which appears to have been the first of its kind, and clearly provides the closest links with Piero's activity as a painter, will be discussed in the next chapter. Here we shall be concerned mainly with the abacus treatise and, to a lesser extent, with the book on the regular solids.

## The abacus treatise

As its brief introductory paragraph tells us, Piero wrote his abacus treatise at the request of friends (or patrons?) and not for use in a school. All the same, the work is very like texts that are known to have been used for teaching in abacus schools. The book on the regular solids is less conventional,

**Fig. 4.1**    Piero della Francesca, *The Baptism of Christ*, tempera on panel, 167.3 × 116.2 cm. National Gallery, London.

being entirely concerned with geometry. It seems to have been written after the abacus treatise, many of whose problems reappear, in a neater or more developed form. The abacus treatise is also somewhat unconventional in its contents. After arithmetic and algebra, it contains an unusually large amount of geometry. In fact, 48 of its 127 pages are concerned with geometry. Many abacus books contained almost no geometry, or dealt only with very simple problems, such as finding the areas of triangles—a type of problem that was very useful in surveying. This was an important matter financially, since much surveying was done with a view to, say, checking the official estimate of the area under cultivation, which was the basis of the landowner's assessment for taxes. Piero's father, Benedetto, was no doubt assessed in this way for the taxes due on his crop of dyer's rocket (*isatis tinctoria*), the plant used to make indigo.

If we are looking for mathematical discoveries, that is new results original to Piero, we must turn to his geometry. However, the earlier sections of the abacus treatise are also of interest. The problems are conventional, and many of them can be traced back through other treatises, some even to the abacus book of Leonardo of Pisa (early thirteenth century, see Chapter 1). In this respect, Piero's treatise is entirely typical of the tradition of abacus mathematics, which was very noticeably traditional and conservative in just this way. So Piero's work gives us a good idea of what was generally taught in abacus schools, and indeed of what he himself was presumably taught in the school he attended. This allows us to place his geometrical work in an appropriate context.

Piero begins not at the very beginning with the writing of Arabic numerals, but with how to multiply and divide fractions. Everything is done by numerical examples. Not all the calculations seem simple enough to be done in the head, though the text gives no indication that other means should be used, say, to divide 188 by 28.

There follows the 'rule of three', whose operation is explained by series of practical examples, starting with one that reminds us that the prosperity of Florence was connected with the cloth trade:

Example. 7 *bracci* of cloth are worth 9 Pounds, what will 5 *bracci* be worth?

The 'pound' mentioned here is a unit of money, derived, centuries earlier, from the value of a pound weight of silver. The Florentine pound was divided into twenty *soldi*, each of which was divided into 12 *denari*. These names are derived from those of corresponding ancient Roman coins. Before decimalisation, in 1971, the British pound was divided in a similar way—into twenty shillings, each further divided into twelve pence—so Piero's calculation bears a strong resemblance to the 'money sums' the present author encountered, and disliked, at junior school. It goes

Do thus: multiply the quantity you want to know by the quantity that the 7 *bracci* of cloth are worth, which is 9 Pounds, thus 5 by 9 makes 45, divide by 7 the result is 6 Pounds with remainder 3 Pounds; make them *soldi*, they give 60, divide by 7, the result is 8 *soldi* with remainder 4 *soldi;* make them *denari*, they give 48, divide by 7, the result is 6 *denari* and $\frac{6}{7}$. So 5 *bracci* of cloth at this price are worth 6 Pounds 8 *soldi* 6 *denari* and $\frac{6}{7}$.

Even without direct memories of childhood, it is easy to recognise the schoolroom tone of this problem. The reader is called by the familiar 'tu' and addressed throughout in the imperative. In fact, the style is a reminder that in abacus schools it was not the pupils who read the text but the master. He would probably read it out loud to his pupils, who would duly copy down what they heard.

Simple problems to be solved by the rule of three are followed by more complicated pieces of arithmetic, each type being provided with a minimal general introduction. Essentially, the instruction proceeds by series of worked examples. Not all the solutions are of a type still familiar in our own time. For instance, we have the following

There is a fish that weighs 60 pounds, the head weighs $\frac{3}{5}$ of the body and the tail weighs $\frac{1}{3}$ of the head. I ask what the body weighs.

The Tuscan pound (weight) contained twelve ounces, each of which was about the same as the modern Imperial ounce, so the Tuscan pound is equivalent to about three quarters of an Imperial pound (that is, about 340 g). All the same, Piero is considering an impressively large fish. In fact, this is one of the problems that can be traced back to the *Abacus book* of Leonardo of Pisa, which was written in Palermo (Sicily) and whose problems are largely Islamic in origin, so the fish could well have been a sea fish rather than a river one. In any case, sixty is a convenient number because it has so many factors. There is, alas, no accompanying picture of the fish.

To the modern eye this problem seems to ask one to use algebra. Piero, however, solves it arithmetically.

Do thus: say that the body weighs 30 pounds, $\frac{3}{5}$ of 30 is 18 which is the head, the tail weighs $\frac{1}{3}$ of the head which is 6. Add together 30 and 18 and 6 they make 54; and you want 60, which is what you have [but] less 6; thus for 30 that I put I got 6 too little. Do the next trial and put the body as weighing 25, the head weighs $\frac{3}{5}$ which is 15, the tail weighs a third of the head, $\frac{1}{3}$ of 15 is 5. Add together 25 and 15 and 5, [which] makes 45; and you want 60, which you have [but] less 15; for the 25 I put I get 15 too little.

That is, Piero has made two guesses at possible answers and has noted by how much the total weight of the fish is mis-estimated in each case. The guess of 30 gives a weight out by 6, the guess of 25 gives a weight out by 15. He then proceeds to use these two incorrect values to calculate the correct solution:

Now multiply 15 by 30 [which] makes 450, then multiply 6 by 25 [which] makes 150; taking this from 450 there remains 300. Take 6 from 15 there remains 9, which is the divisor. Divide 300 by 9 there results $33\frac{1}{3}$, take $\frac{3}{5}$ of $33\frac{1}{3}$ [which] are 20, and one third of 20 is $6\frac{2}{3}$. Accordingly the body weighs $33\frac{1}{3}$, the head weighs 20, the tail weighs $6\frac{2}{3}$; which added together give 60, as I said the fish weighed.

Piero has called this method the *regola de la positione*, literally 'rule of position'. Historians usually call it the 'rule of false position'. The word 'position' is, however, being used here in a rather strange sense, to mean something like 'substitution'—and in my translation, above, Piero's word 'positione' was in fact rendered by 'trial'. The linguistic problem is, of course, a pointer to the fact that this method was first applied in Arabic.

Piero nowhere explains how the method works. For a present-day reader the simplest way of explaining is probably to use algebra (and to use it in a form which Piero would have regarded as entirely outlandish). Since all weights are given in proportions, it is clear that the estimate of the whole weight of the fish will be proportional to the weight guessed for the body. That is, if we guess the body weighs $x_1$, the whole fish is estimated as weighing, say, $kx_1$, and if we guess $x_2$ then the total weight is $kx_2$. Let the actual weight of the fish be A and the remainders from the guesses $r_1$ and $r_2$ respectively, then we have the two equations

$$kx_1 + r_1 = A \qquad (1)$$
$$kx_2 + r_2 = A \qquad (2)$$

The value we want to know is $k$, or rather $A/k$. So we eliminate the second term on the left of each equation (that is the terms in $r_1$ and $r_2$) by multiplying (1) by $r_2$ and (2) by $r_1$ and subtracting, which gives us

$$k(r_2 x_1 - r_1 x_2) = A(r_2 - r_1) \tag{3}$$

Dividing through appropriately gives

$$A/k = (r_2 x_1 - r_1 x_2)/(r_2 - r_1) \tag{4}$$

That is, the number we require is a fraction whose numerator is the product of the second remainder by the first guess (15 times 30 for Piero's example) minus the product of the first remainder by the second guess (6 times 25 for Piero), and whose denominator is the second remainder minus the first (15 minus 9 for Piero). If one wishes to put the matter in geometrical terms one can say that Piero has solved a linear problem by finding two points on the line. What he has got, from Leonardo of Pisa and his Islamic sources, is a general method of solving linear problems. To a reader familiar with algebra—and we may note that Piero has not yet mentioned algebra—it is clear that we could have evaluated $k$ directly. So Piero's traditional arithmetical method looks clumsy. However, it has the advantage of allowing one to choose to work with simple numbers, that is to avoid having to deal with fractions in the course of the main part of the working. Division comes only at the end.

After arithmetic, Piero turns to algebra, which he calls 'algibra'. He says it deals with fractions and whole numbers and squares and cubes of numbers. A whole number is also called a root (*radice*) and a square number (*quadrato*) is also called a power (*censo*). He then gives the six 'rules' (*regole*) of algebra. A 'rule' in this context is what would now be called a type of equation. However, Piero does not actually write out equations as such. Everything is done in ordinary sentences, with no attempt at tabulation. In the later fifteenth and sixteenth centuries abbreviations began to be introduced, resulting in algebra which looks much like a knitting pattern, and is about as self-explanatory to the uninitiated. Piero's algebra, however, looks like his arithmetic, that is ordinary prose dotted with numbers (some of them fractional). His prose style is not particularly elegant. My translations attempt to copy it as closely as possible.

Piero tells us that the first three 'rules' of algebra are simple, the second three composite.

The three simple rules are when in arithmetical or geometrical questions the thing or roots are equal to a number, or the powers are equal to the things, or the power is equal to a number.

Piero's word for the unknown is '*cosa*', literally 'thing'. Since vernacular spelling is phonetic, the word is sometimes written '*chosa*', to give the aspirated Tuscan pronunciation. The word 'thing' is quite often omitted, though once we come to numerical examples the fact that a number specifies a number of 'things' is indicated by putting a bar over it, and for a number of 'powers' a small square. This style of writing equations out in words—which is sometimes called 'rhetorical algebra'—makes it clear why Piero needs to say, as he at once does say, that the second rule (powers equal to things) is equivalent to the first (thing equal to number). He then goes on to explain how to solve for the thing in each rule. This is done in the abstract, without recourse to numerical examples.

There follow the three composite rules:

And the composite [rules] are when the powers and the things are equal to the numbers, and when the powers and the numbers are equal to the things, and when the power is equal to the things and the number.

These three types, it will be noted, all involve only addition, not subtraction. They are (ultimately) derived from Euclid's *Elements*, where quadratic equations were solved by considering the addition of areas. The concept of a negative number was not to be introduced until late in the sixteenth century. For Piero and his contemporaries, in doing algebra, a number meant a positive integer, a rational fraction (that is a fraction expressible in the form *a/b* where both *a* and *b* are positive integers), or a root of this kind of number. There was, however, room for argument as to whether a root should really be considered as a number, since the word '*numero*' (after the Latin *numerus*) was recognised in some contexts as being a translation of the Greek 'arithmos' ($\dot{\alpha}\rho\iota\theta\mu\dot{o}\varsigma$), which meant a positive integer. All the same, a root or an arithmetical expression involving a root was acceptable as an answer to a problem, and in practical matters that was, of course, what was important.

After an abstract account of how to solve for the thing in each of the composite rules, Piero passes on to numerical examples and then to problems posed as practical. Here we again encounter the fish, though with its weight and proportions slightly modified.

A fish weighs 50 pounds, the head weighs $\frac{1}{3}$ of the body, the tail weighs $\frac{1}{4}$ of the head. I ask what the body weighs, and what the head weighs and what the tail weighs.

This time the solution is algebraic.

Put the weight of the body as $\overline{1}$ thing, the head weighs $\frac{\overline{1}}{3}$, the $\overline{1}$ tail weighs $\frac{\overline{1}}{4}$ of $\frac{1}{3}$; and $\frac{1}{3}$ and $\frac{1}{4}$ of a third are $\frac{5}{12}$, so you have $\overline{1}\frac{5}{12}$ equal to 50. Make into twelfths [and] you will have $\overline{17}$ of them equal to the number 600; divide by 17 the result is $35\frac{5}{17}$, this is the value of the thing we put as the weight of the body, [so] the body weighs $35\frac{5}{17}$. The head weighs $\frac{1}{3}$ which is $11\frac{13}{17}$, the tail weighs $\frac{1}{4}$ of the head which is $\frac{1}{12}$ of the thing whose value is $2\frac{16}{17}$, so the weight of the tail is that: $2\frac{16}{17}$.

Piero does not actually check that the sum of these parts is 50. Presumably the pupil is by now considered too advanced to need such prompting. Nor does Piero compare this algebraic method with the arithmetical method of false position. No doubt then as now an attentive pupil would notice that depending on how the problem was set it would be more or less awkward to be compelled to work with the numbers given, as in the algebraic solution, rather than being able to choose some of one's own numbers, as in the method of false position. In any case, the fish problem is only one of the problems previously solved by arithmetic which Piero now proceeds to solve by algebra. As far as Piero is concerned, when we have relatively easy problems, arithmetic and algebra clearly coexist as alternative approaches. In fact, well into the sixteenth century algebra was generally regarded as no more than an extension of arithmetic. Like arithmetic, it dealt with numbers. All the same, Piero's later examples show that for more difficult problems algebra has the edge as a means of solution.

A few of Piero's more advanced algebraic examples are numerical versions of geometrical problems. For instance, his very last example is

There is a triangle, which has base 12 and its height is 10, and the two other sides add up to 24. I ask what each is.

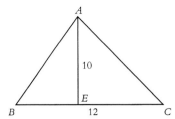

**Fig. 4.2** The figure for Piero's final algebraic problem. In modern notation, we are given *BC* = 12, *AE* = 10, *AB* + *AC* = 24, and the problem is to find *AB* and *AC*.

The triangle, with Piero's lettering, is shown in our Fig. 4.2. Piero does not supply a diagram. The problem gives rise to a quadratic equation—though some of Piero's earlier examples gave him cubics—and the solution is that *AC* is equal to '12 plus the root of $2\frac{2}{3}$' and *AB* is '12 minus the root of $2\frac{2}{3}$'.

## Geometry in the abacus treatise

Perhaps this geometrical problem was intended to provide a smooth transition to the geometrical problems that follow in the next section, but Piero in fact provides something of a jolt by reverting to the simplest abacus book types of geometrical problem, namely finding sides, heights and areas of triangles. The first problem is

Let the triangle *ABC* have equal sides, and let each be 10 *bracci*. I ask what is its height.

Piero consistently uses '*bracci*' as the plural form of *braccio*, though, as we mentioned in Chapter 3, the usual plural was *braccia*. In vernacular texts of this period grammatical rules, and technical terms, have a do-it-yourself element that is rather disconcerting to anyone accustomed to today's habit of being careful about definitions. The solution to the problem about the triangle begins in the best schoolmasterly manner 'There are many ways [of solving this problem], I shall show you two or three.' In fact Piero first solves the problem for the equilateral triangle and then moves on to the abacists' favourite scalene triangle, whose sides are 15, 14 and 13. Its height, above the side of length 14, is 12 (see Appendix). Piero's solutions are entirely numerical. Like the algebraic solution we have just mentioned, they make repeated use of Pythagoras' Theorem, but do not say so. Indeed, Piero does not even remark that the triangles concerned are right-angled. This is the standard style of abacus book geometry, and, as we shall see, it is rather different from what we find in Piero's original contributions, which concern three-dimensional figures.

The first three-dimensional figure to appear is the regular tetrahedron, one of the five regular solids that had been known since Antiquity: the others being the octahedron, the icosahedron, the cube and the dodecahedron (see Fig. 4.3). Before this, Piero has worked through the properties of triangles, squares, rectangles and various regular polygons. He has provided what is almost a summary of the corresponding parts of Euclid's *Elements*, but has done so in the form of numerical examples and without formal proofs. What we have is Euclid made useful. (In the sixteenth century, books written in this style tended to have titles like *Practical geometry*.) With the appearance of solid figures we have reached *Elements* Book 13. This is far beyond the scope of the standard

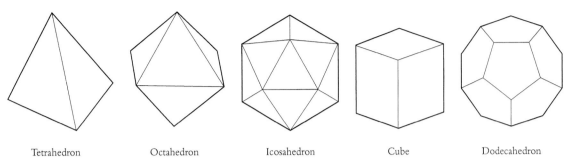

|  |  |  |  |  |
|---|---|---|---|---|
| Tetrahedron | Octahedron | Icosahedron | Cube | Dodecahedron |

**Fig. 4.3**  The five convex regular solids (the 'Platonic solids').

geometry taught in universities in Piero's time, though a very brief mention of the five regular solids was to be expected in connection with Plato's identification of them with the five 'elements' (earth, water, air, fire and 'æther', the element of the heavens) in his dialogue *Timæus*. *Timæus* was the one dialogue of Plato that was known and studied throughout the Middle Ages, and it is also the one which Raphael (1483–1520) chose to show to identify Plato in his famous picture *The School of Athens* (painted in fresco in the Vatican in about 1509). Because of their importance in *Timæus*, the five regular bodies are sometimes known as the Platonic solids.

Plato is concerned to draw analogies between the properties of the regular polyhedra and those of the corresponding 'elements'. For instance, the transmutation of water into air is analogous to the breaking up of icosahedra, each into its twenty triangular faces, which then recombine to form octahedra (each with eight triangular faces). The mathematical properties concerned are largely very simple ones. Euclid, and Piero, are concerned with properties which have more interesting mathematical implications. For instance, one of Piero's problems on the tetrahedron is

There is a spherical body whose diameter is 7; I want to put in it a figure with 4 equilateral triangular faces, so that the corners [of the figure] touch the circumference [of the sphere]. I ask what its edges are.

The choice of diameter points back to an earlier problem in which Piero has used Archimedes' value of $\pi$, namely $\frac{22}{7}$, to find the surface area of the sphere. Piero duly gives Archimedes credit for this, which presumably indicates that Archimedes' theorems on the sphere were considered less well known than Pythagoras' theorem on the right-angled triangle.

After the analogous problem of finding the size of a cube inscribed in a sphere, Piero turns to a solid not mentioned in Euclid (and to which no name is given):

There is a spherical body whose diameter is 6; I want to put in it a body with 8 faces, 4 triangular and 4 hexagons. I ask what its edge is.

This body is now called a truncated tetrahedron. It is one of the set of polyhedra known as the Archimedean solids. Like the five regular bodies considered by Euclid, these solids are convex and have regular faces which meet in the same way at every corner of the solid, but the faces of each Archimedean solid are not all of the same kind. There are thirteen such solids. They owe their present day names to Johannes Kepler (1571–1630), who described, and illustrated, all thirteen of them in the second book of his *Five books of the harmony of the world* (*Harmonices mundi libri V*, Linz, 1619); see Fig. 4.4. The solids are called Archimedean because the Alexandrian mathematician Pappus, writing

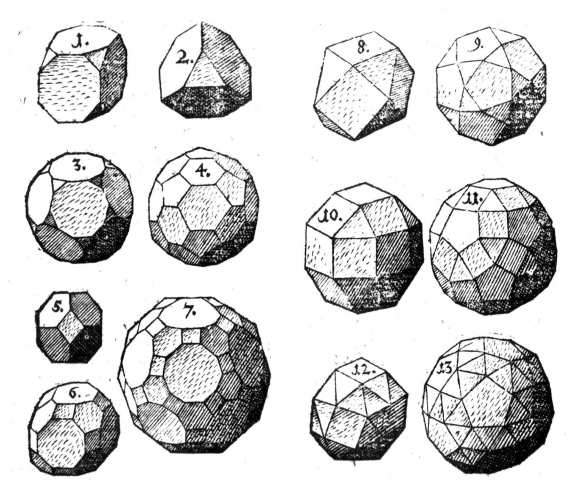

**Fig. 4.4** The thirteen Archimedean solids, from Johannes Kepler, *Five books of the harmony of the world*, 1619, Book 2, Proposition 28. The illustrations were drawn by Kepler's friend the Professor of Mathematics at the University of Tübingen, Wilhelm Schickard (1592–1635). The modern names for the solids shown here are 1. truncated cube, 2. truncated tetrahedron, 3. truncated dodecahedron, 4. truncated icosahedron, 5. truncated octahedron, 6. truncated cuboctahedron, 7. truncated icosidodecahedron, 8. cuboctahedron, 9. icosidodecahedron, 10. rhombicuboctahedron, 11. rhombicosidodecahedron, 12. snub cube, 13. snub dodecahedron.

in the fourth century AD, ascribes their discovery to Archimedes about six centuries earlier. However, Pappus' descriptions of the solids in question are entirely in the style we have quoted from Piero. That is, he lists the numbers and types of faces of each solid, but gives no indication of how they fit together to form a three-dimensional structure. Piero does: not in his text, except by implication, but in his diagrams (e.g. Fig. 4.5). Piero is thus, entirely reasonably, credited with having rediscovered the five Archimedean solids he describes. Two of them are found in the abacus treatise, namely the truncated tetrahedron and the cuboctahedron (see below). The truncated tetrahedron reappears in the book on the five regular solids, together with the truncated cube, the truncated octahedron, the truncated dodecahedron and the truncated icosahedron (for the forms of these solids see Fig. 4.4). It is not

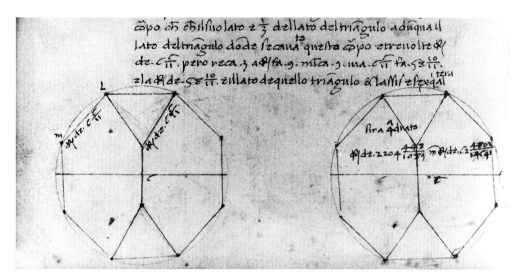

**Fig. 4.5**   Piero della Francesca's illustrations of the truncated tetrahedron, from his *Trattato d'abaco* (Florence, Biblioteca Medicea-Laurenziana, Codice Ashburnhamiano 280 (359*)).

clear whether Piero knew Pappus' descriptions of these solids. Probably his not mentioning the name of Archimedes should be taken as an indication that he did not.

Having found the side and the volume of the truncated tetrahedron, Piero next turns to the solid we now call the cuboctahedron.

There is a spherical body, whose diameter is 6 *bracci*; I want to put in it a figure with fourteen faces, 6 square and 8 triangular, with equal edges. I ask what each edge will be.

The accompanying diagram is shown in Fig. 4.6. As for the truncated tetrahedron, Piero explains that this solid also is formed by cutting corners off one of the regular solids:

Such a figure as this is cut out from the cube, because it [the cube] has 6 faces and 8 corners; which, cutting off its 8 corners, makes 14 faces, that is thus. You have the cube *ABCD.EFGH*, divide each side in half: *AB* in the point *I*, and *CD* in the point *L*, *BD* in the point *K*, *AC* in the point *M*, ... *EG* in the point *V*.

Piero's diagram, our Fig. 4.6, does not show the original cube, but only the final solid obtained by cutting off its corners. If, assuming the role of pupil, we attempt to supply the complete diagram for ourselves, it becomes clear that Piero's conventions for lettering diagrams are not those usual in the twentieth century. The edges of the cube, which are listed for bisection, are *AB*, *CD*, *BD*, *AC*, ... , so it is clear that one face must be lettered as in Fig. 4.7. That is Piero has lettered the square left to right in rows, but the points of bisection are lettered anticlockwise in alphabetical order. Comparison with Piero's own diagram, Fig. 4.6, shows that Fig. 4.7 corresponds to the top face of the cube. Further reference to the list of edges indicates that the diagram of the cube with the points of bisection of its edges should be as in Fig. 4.8.

Piero's proposition continues

Draw a line from *T* to *P* passing through the centre *K*, which [line] is in the power equal to [the sum of] the powers of the two lines *TN* and *NP* because *N* is a right angle opposite the line *TP*; ...

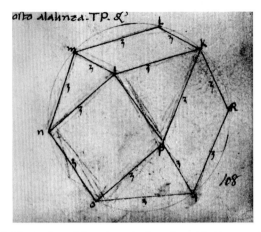

**Fig. 4.6** Piero della Francesca's illustration of the cuboctahedron, from his *Trattato d'abaco* (Florence, Biblioteca Medicea-Laurenziana, Codice Ashburnhamiano 280 (359*)).

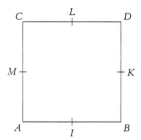

**Fig. 4.7** One face of Piero's cube, with edges bisected to form the vertices of a cuboctahedron.

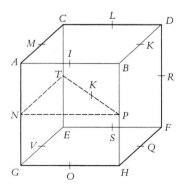

**Fig. 4.8** Cube for Piero's construction of the cuboctahedron. The point $K$ on the line $TP$ is the centre of the given sphere.

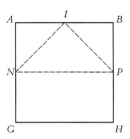

**Fig. 4.9** Face of the cube from Fig. 4.8.

The *K* concerned is, clearly, the centre of the sphere (or the cube) and is not the same as the *K* constructed as the midpoint of the line *BD*. As we shall see later, in connection with a theorem in Piero's perspective treatise, using letters twice in one diagram seems not to have been so thoroughly forbidden in Piero's day as it is in ours. The triangle *TNP* with which he is concerned is shown in broken lines in Fig. 4.8.

It will be noted that here Piero is using Pythagoras' Theorem, with due reference to the triangle being right-angled. He goes on to point out that *TP* is a diameter of the sphere, and therefore of length 6, so that its power is 36 and the powers of *NP* and *NT* are therefore each 18, making each line √18. He then turns to the triangle *INP*, shown in our Fig. 4.9, omitting to point out that it has a right angle at *I*, but using Pythagoras' Theorem again, to obtain the result that each of the lines *NI* and *IP* is √9:

So that I say that the side of the figure with 14 faces, 6 square and 8 triangular, is the root of 9, which is 3.

Piero's method of working, by considering plane figures within the solid one, makes his treatment of the cuboctahedron like Euclid's treatment of the regular solids in *Elements* Book 13. One could, in fact, dispense with a diagram of the solid figure. In supplying one, Piero is inviting us to visualise what the solid will actually look like. So far so painterly. It is not difficult to see Piero's power of visualising what one would obtain by cutting the corners off a cube, and thus discovering (or rediscovering) the cuboctahedron, as being connected with his skill as an artist.

However, closer inspection shows that, solid-looking though it is, Piero's diagram of the cuboctahedron is not drawn in correct perspective. If we consider the vertices *M*, *N*, *O*, *Q*, *R*, *L*—all of which are shown in Piero's diagram (Fig. 4.6) and my own (Fig. 4.8)—it is clear that they form what one might call an equator of the solid, a regular hexagon whose vertices are the ends of the edges radiating from the vertices of the triangle *IPK*. This hexagon in fact gives us another way of finding the edge of the solid. The circumcircle of the hexagon is obviously a great circle of the circumsphere of the solid, so the side of the hexagon (which is the edge of the solid) is equal to the radius of the sphere; that is, in Piero's example, it is 3. Now, the circumcircle of the equatorial hexagon is, apparently, shown in Piero's diagram, but the point *K* (the midpoint of the edge *BD* of the cube) is also shown as lying on the curve. This suggests that we may be being shown the body as it would appear if it were rotated about the edge *LR* until *K* was seen against the arc *LR*. That would make the curve through *M*, *N*, *O*, *Q*, *R*, *L* and *K* a tilted view of the equatorial circle, that is an ellipse. The diagram in the manuscript (which is believed to be in Piero's hand) is too small

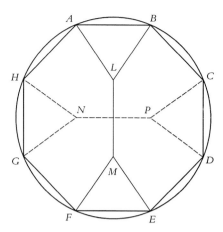

**Fig. 4.10** Copy of Piero's diagrams of the truncated tetrahedron, with added lettering and dashed lines to shown hidden edges.

for one to check the shape of the curve, but close inspection of the paper shows the mark of a compass point in the appropriate position. The curve is a circle. So the circle stands for the sphere, and rather than draw his figure in correct perspective Piero has chosen to emphasise the relation of the vertices to the sphere by making as many of them as he can lie on the circle.

This realisation must, of course, make us look again at Piero's diagrams of the truncated tetrahedron (Fig. 4.5). Close inspection of the manuscript page again shows the mark of a compass point in the appropriate place, indicating that the outer curve is indeed a circle. An approximate copy of Piero's diagram is shown in Fig. 4.10, in which lettering has been added as well as dashed lines to indicate hidden edges. From considerations of symmetry, it is clear that the points *A*, *B*, *E*, *F* lie in the same plane and form the vertices of a rectangle. Moreover, the points *C*, *D*, *G*, *H* are the vertices of a rectangle of similar shape and size, lying in a plane parallel to that containing *A*, *B*, *E*, *F*. Thus, in space, the points *A*, *B*, *E*, *F* are concyclic, and the circle on which they lie is the same size as that on which the points *C*, *D*, *G*, *H* are located. So if Fig. 4.10 is taken to be a plan of the truncated tetrahedron—an orthogonal projection onto a plane parallel to those of the two rectangles concerned—then we should see all eight vertices *A*, *B*, *C*, *D*, *E*, *F*, *G*, *H* lying on a single circle. So our diagram, which in this respect is a copy of Piero's, could be read as a plan of the solid. However, in that case the circumcircle of the octagon *ABCDEFGH* is not a great circle of the circumsphere of the solid. However, its significance is probably the same as that of the circumcircle in Piero's diagram of the cuboctahedron (Fig. 4.6), where the circle could be a great circle of the circumsphere. That is, the circle in the figure of the truncated tetrahedron is likewise intended to convey the information that the vertices of the solid lie on a sphere. Consistency between the conventions used in successive drawings seems not to have been of overriding importance.

A further inconsistency appears if we look at Piero's diagram of the regular dodecahedron, shown in Fig. 4.11. Here, close inspection of the actual page again shows the mark of a compass point, so the outer curve is a circle. But several of the outermost vertices of the solid do not lie on it. The actual drawing of the solid is convincingly three dimensional, suggesting that it may be in correct perspective—which would, of course, account for the vertices failing to lie on the circle. Confirmation of this is provided by the fact that in the drawing the vertices of the solid

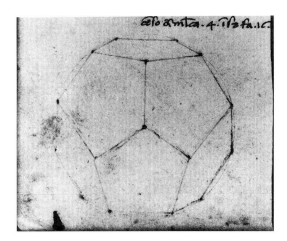

**Fig. 4.11** Piero della Francesca's illustration of the regular dodecahedron, from his *Trattato d'abaco* (Florence, Biblioteca Medicea-Laurenziana, Codice Ashburnhamiano 280 (359*)).

are marked with indentations, apparently caused by a blunt point being pressed against the paper, which (as is usual with paper of this date) has a rather spongy texture. Indentations like these were commonly used by artists for the transfer of drawings, so it seems likely that Piero had a separate correct perspective drawing of the dodecahedron which was then used to produce this drawing in his treatise.

It would seem that for the regular dodecahedron Piero decided that correct naturalistic perspective provided his best means of communication, but that in showing the two Archimedean polyhedra he made different decisions. The regular dodecahedron is treated in *Elements* Book 13, where it is constructed with its vertices lying on a sphere. So Piero may have felt his readers would know about the sphere and would not be troubled by the apparent vagueness of its relation to the solid in his diagram. He certainly could not rely upon this kind of prior knowledge for the two Archimedean solids. This may account for his having decided to draw his diagrams in such a way as to emphasise that the vertices of these solids lay on a sphere. In the case of the cuboctahedron his decision involved a departure from correct perspective. In the case of the truncated tetrahedron it led at least to a drawing of a circle which in more normal circumstances would not have been read as a representation of the sphere concerned.

The non-naturalistic style of some of Piero's diagrams may seem to divide his work as a mathematician from his work as a painter. In fact, if we look at the way three-dimensional figures were usually drawn in mathematical texts of his time, we can see that Piero's illustrations are unusual, indeed apparently highly innovative, in showing so great a degree of naturalism. It was, for instance, usual to draw a cone as a circle combined with a pair of lines, a cylinder as two circles linked by two parallel lines, a square pyramid as a pencil of four lines radiating to the vertices of a square, and so on (see Fig. 4.12). There do not seem to have been any very definite conventions that governed the drawing of more elaborate figures. In any case, the purpose of such diagrams seems to be to show which points were joined by which edges. The diagrams appear not to be aids to visualisation in a more general sense. By contrast, Piero's diagrams give a much more naturalistic impression of the actual appearance of the solid concerned.

 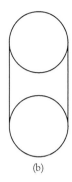 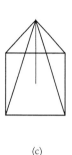 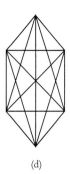

(a)  (b)  (c)  (d)

**Fig. 4.12**  Three-dimensional geometrical figures, in the style in which they were usually drawn in the fifteenth century: (a) cone, (b) cylinder, (c) pyramid on square base, (d) regular octahedron.

## The short book on the five regular solids

The same variety of styles for drawing solids that we find in the abacus treatise is also found in the single known copy of Piero's *Short book on the five regular solids* (*Libellus de quinque corporibus regularibus*). It is not known who drew these illustrations. The text of the book is in Latin—as one might have deduced from the title, but for the fact that Piero's treatise on perspective also has a Latin title but was (it seems) written in the vernacular. The book on the regular solids may also have originally been written in the vernacular, that is in Piero's native Tuscan dialect. In any case, almost all of its content is directly taken from the abacus treatise, though the later work presents some problems in a neater or more developed form—which, as we have already pointed out, is good reason for supposing the book on the regular solids to have been written after the abacus treatise.

The preface to the book on the regular solids tells us that it was written as a companion piece to Piero's treatise on perspective, and we know the works were at one time shelved next to one another in the library in the Ducal Palace at Urbino. Piero had worked in Urbino—as a painter, that is (see Fig. 4.13)—so it is not surprising that the Ducal library should contain a copy of the perspective treatise. The book on the regular solids was dedicated to the Duke of Urbino, and may possibly have been written at his request. It is not hard to imagine that someone who knew the abacus treatise might recognise that its geometrical work was unusually interesting and suggest it could form the basis for an independent treatise. In any case, Piero's book on the regular solids is, it seems, the first treatise since ancient times to be devoted entirely to geometry and the presentation of new geometrical results, so its origin does stand in need of some sort of explanation.

All the same, hailing the book on the regular solids as any kind of 'first' tends to jar somewhat with the impression one receives from the actual text. Apart from being in Latin, the work reads like an abacus book. As in his abacus treatise, Piero proceeds by means of worked numerical examples, with very little linking text. As in the abacus treatise, the new Archimedean solids are introduced by means of problems involving inscribing them in a sphere. And when, as the penultimate problem of the work, Piero proposes finding the volume of a very irregularly shaped body, 'such as the statue of a person or an animal', his solution is to instruct the reader to construct a wooden tube, of appropriate dimensions, with square cross-section, made of stout planks, with its corners well sealed, fill it two thirds full of water, and so on. The volume of the statue is found by the

**Fig. 4.13**  Piero della Francesca, *Federigo da Montefeltro and his wife Battista Sforza*, each 47 × 33 cm, on panel, Galleria degli Uffizi, Florence. Piero's *Short book on the five regular solids* was dedicated to their son Guidobaldo.

volume of water displaced. One can hardly help suspecting that the many historians who have described this treatise as 'Euclidean' (presumably on account of its being exclusively concerned with geometry) have not actually read as far as this passage.

As we have mentioned, the abacus book style of the treatise on regular solids extends to Piero's presentation of his Archimedean solids, which he confusingly describes as 'irregular'. However, a general idea seems to lie behind the series of numerical examples. The five solids presented (numbers 1 to 5 in Fig. 4.4) are all obtained by the method used to generate the truncated tetrahedron in Piero's *Abacus treatise*, namely cutting off the corners of a regular solid so that each of the regular $n$-gon faces becomes a regular $2n$-gon. Figure 4.14 shows the effect on the triangular face of a regular tetrahedron, with lettering corresponding to that in Fig. 4.10. In contrast, in the method of truncation used to produce a cuboctahedron, the cut is made to the centre of each edge of the face of the cube (see Fig. 4.7). This method turns a regular $n$-gon into another, smaller, regular $n$-gon in a different orientation. Since Piero's *Abacus treatise* includes an example of each kind of truncation, while his book on the regular solids gives all five solids obtained by the first kind, he must surely be recognised as having distinguished between the two processes. He has invented the notion of truncation in its modern mathematical sense.

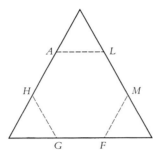

**Fig. 4.14** The face of a regular tetrahedron subjected to truncation. This is the type of truncation now usually known as 'semi-truncation'. (Lettering as in Fig. 4.10.)

Piero's *Abacus treatise* never found its way into print under its author's own name in the Renaissance, though almost all its arithmetical and algebraic problems are to be found in a hugely successful textbook written by Luca Pacioli (*c.* 1445–1514): *Summary of arithmetic, geometry, proportions and proportionality* (*Summa de arithmetica, geometria, proportioni e proportionalità*, Venice, 1494). This publication is one piece of evidence that Pacioli had access to Piero's manuscripts (though possibly only after his death in 1492). Another piece of evidence pointing in the same direction is the appearance of Piero's *Short book on the five regular solids* as the third section of Pacioli's *On divine proportion* (*De divina proportione*, Venice, 1509), with a dedication that implies the piece was written by Pacioli himself. The text, like that of the rest of *De divina proportione*, is in Italian, with Venetian dialect spelling that was presumably supplied by the typographers. Vasari, writing about forty years later, was to complain of Pacioli's conduct in printing Piero's work under his own name. By then, it was usual to recognise that there was such a thing as intellectual property. Indeed whether such property could be inherited, in the manner of a second-best bed, was one of the issues underlying the highly flamboyant series of charges and counter charges associated with the battle over the invention of the general solution to cubic equations in the mid-sixteenth century: if Niccolò Tartaglia (*c.* 1499–1554) had filched the solution from the legitimate heirs of its discoverer Scipione dal Ferro (1465–1526) then how could he complain when it was published by someone else? Tartaglia in any case denied dishonesty, claiming independent invention following his victory in an equation-solving contest with one of Scipione's pupils. (For the sequel see Chapter 6.) In Pacioli's time intellectual property rights were vaguer. *On divine proportion* is chiefly famous for the illustrations drawn for its first part by Pacioli's friend Leonardo da Vinci (1452–1519). For this part of his work Pacioli has used Piero's abacus treatise, but has also added one new Archimedean solid that he seems to have discovered for himself (or perhaps with some help from Leonardo?), the solid now called the rhombicuboctahedron (see Fig. 4.15). The complicated story of the Renaissance rediscovery of the Archimedean solids in fact involves other people notable for their interest in art as well as mathematics, namely Albrecht Dürer (1471–1527) and Daniele Barbaro (1513–1570) (both discussed in Chapter 6). Johannes Kepler, the professional mathematician who tidied everything up in the best *Deus ex machina* way, makes no claim to originality, but also gives no references to any predecessor after Pappus. The intricacies of the rediscovery are not relevant here, but it is surely of interest that painters, using their skills at visualising structures in three dimensions, made a significant contribution to the revival of a piece of geometry that was recognised as having its place in the learned tradition.

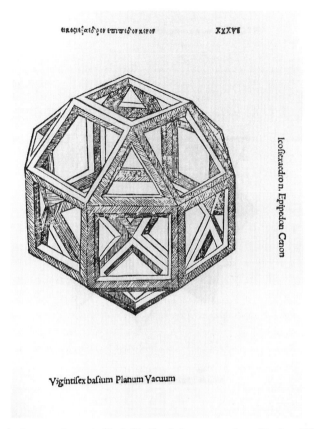

Fig. 4.15 Rhombicuboctahedron, as shown in Pacioli's *De divina proportione* (Venice, 1509), woodcut based on a drawing by Leonardo da Vinci.

The final problem in Piero's *Short book on the five regular solids* returns to two-dimensional geometry. It is a rather cumbersome affair involving a triangle, two of whose sides are constructed as tangents to a circle touching the third side. As the length of this third side is 14 and the diameter of the circle is 8, readers may perhaps guess that we are once again concerned with the abacists' favourite triangle, whose sides are 13, 14 and 15 (on which see Appendix). For all the entirely geometrical nature of the content of the book on the regular solids, its style and its problems show its closeness to the practical mathematics of the abacus schools. As we shall see in the next chapter, the same is true, though in a rather different way, of the work to which the book on the regular solids was proposed as a companion, namely Piero's treatise on perspective. In studying that work, however, it becomes relevant that we are dealing not only with a skilful mathematician but also with a first class painter.

# Chapter 5

## PIERO DELLA FRANCESCA'S PERSPECTIVE TREATISE

Piero della Francesca's perspective treatise is the first of its kind—as far as we know and, it would seem, as far as Piero knew also. It is called *On perspective for painting* to make clear that we are concerned not with ordinary natural optics, which at this time was sometimes known as 'common perspective' (*perspectiva communis*), but with the special kind used by painters. As we shall see, Piero takes care to show that this new part of perspective should be seen as a legitimate extension of the older established science.

*On perspective for painting* survives in fifteenth-century manuscript versions both in Piero's vernacular Tuscan and also in Latin. The Latin is clearly a translation. However, even manuscripts of the Tuscan original have the Latin title *De prospectiva pingendi*. Piero's style is, none the less, largely that of the abacus school, and that of his own abacus treatise. The reader is called 'tu' and is addressed in the imperative. Instruction proceeds almost exclusively through series of worked examples. At first glance, the treatise seems to consist entirely of orders in the form 'draw the line *AB*'. Its grim repetitiousness contrasts with the delicate neatness of the illustrations which are provided at the end of every proposition. In effect, almost all the propositions could be read as telling the apprentice how to make a copy of the diagram supplied at the end. This style reflects that of artists' workshops, where the apprentice learned to draw by copying series of drawings by his master. Such drawings were preserved in the form of manuals available in the workshop. They provided a repertory of stock images whose repeated use can sometimes help to identify the workshop in which a particular painting was produced. This does not apply only to hack painters. Elegant renderings of a heron-like bird in paintings by Giovanni Bellini and his brother-in-law Andrea Mantegna (1431–1506) are derived from a drawing by Giovanni's father, Jacopo Bellini (*c*. 1400–1470/1), in whose workshop they had both been apprentices.

The detailed drawing instructions given in Piero's treatise, and the provision of a finished drawing at the end of each proposition, thus suggest that Piero saw himself as essentially providing a workshop manual to teach the apprentice to draw in perspective. Since the 'solution' to each problem is a drawing, the perspective treatise has a purely geometrical look that makes parts of it appear closer to the *Elements* than Piero's other works do. We do not find geometrical problems posed in numerical terms, as in the abacus treatise and the book on the regular solids, though Piero does occasionally put in a numerical example, apparently as some kind of check (for those feeling confused by his geometrical reasoning?). Despite the Euclidean look of the work, Piero makes it clear in his introduction to the first book that his concern is not with geometry as such.

A point is that which has no parts, accordingly the geometers say it is [merely] imagined (*inmaginativo*); the line they say has length without width.

And because these are not seen except by the intellect and I say I am dealing with perspective in demonstrations that are to be taken in by the eye, on this account it is necessary to give a different definition. So I shall say the point is a thing as tiny as it is possible for the eye to take in; the line I say is extension from one point to another, its width being of the same nature as that of the point. A surface I say is width and length enclosed by lines. Surfaces are of many kinds (*ragioni*), such as a triangle, a quadrangle, a tetragon, a pentagon, a hexagon, an octagon and with more and different edges, as will be shown in the figures.

These are the last two paragraphs of Piero's introduction, and it will be noted that although they indicate that his purpose is essentially practical they do also, at the end, lapse into Greek (though in Latin script). This is a characteristic habit of humanists of the time. In this case, the resonances are Euclidean. The 'tetragon' shown in Piero's diagram is in fact a rectangle. Unlike some other writers of elementary texts—for instance Albrecht Dürer—Piero does not concern himself with making the very simple distinction between straight lines and curved ones. For Piero a line is straight. It is also, as was usual, finite, being drawn from one point to another. Similarly, surfaces are conceived as having edges and edges as defining surfaces. So a 'plane' is a flat shape with a defined boundary, whose nature (say, triangular or square) needs to be specified; and a 'triangle' is a triangular area not a configuration of three lines.

Piero divides his treatise into three books. Each has a short discursive introduction. The first book begins with optical and mathematical preliminaries before setting out a series of problems concerning the drawing of plane figures. The second book uses some of these figures as ground plans, and essentially deals with prisms. The third book deals with more complicated shapes—and uses a method different from that of the previous two books.

## Piero's preliminaries

The introduction to Piero's first book makes it clear that his treatise is only about perspective, not about painting as a whole or even about everything the painter needs to know. The introduction begins

Painting has three principal parts, which we say are drawing (*disegno*), proportion (*commensuratio*) and colouring (*colorare*). Drawing we understand as meaning outlines and contours contained in things. Proportion we say is these outlines and contours positioned in proportion in their places. Colouring we mean as giving the colours as they are shown in the things, light and dark according as the light makes them vary. Of the three parts I intend to deal only with proportion, which we call perspective, mixing in with it some part of drawing, because without this perspective cannot be shown in action; colouring we shall leave out, and we shall deal with that part which can be shown by means of lines, angles and proportions, speaking of points, lines, surfaces and bodies.

The words come to life when we look at Piero's paintings. The *Baptism* (Fig. 4.1) and the portraits of Federigo da Montefeltro and Battista Sforza (Fig. 4.13) show Piero's characteristic delicacy in handling the fall of light, his care over reflections, and particularly in the *Baptism*, the use of scaling to achieve a sense of space. Though some mathematical thought must certainly have gone into the construction of the *Baptism*, it has not left traces like those in, say, Uccello's *St George and the dragon* (Fig. 2.13).

A particularly striking instance of Piero's showing how light makes colour vary can be seen in Fig. 5.1, which shows one of the scenes in the fresco cycle *The Story of the True Cross*, in the

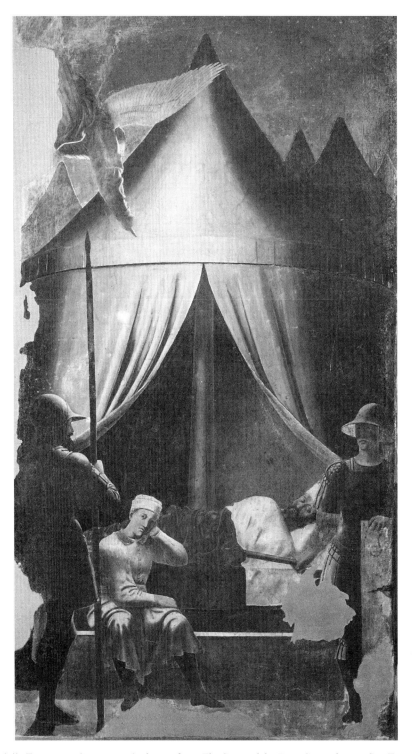

**Fig. 5.1**  Piero della Francesca, *Constantine's dream*, from *The Story of the True Cross*, fresco, San Francesco, Arezzo.

church of San Francesco, Arezzo. The figures are life size. The scene is that of Constantine's dream. On the eve of a battle, the Emperor, who was not yet a Christian, dreamed that an angel presented him with a cross bearing the legend 'In this sign you will conquer'. Piero has left out the words, presumably trusting to our making the connection with the cross held in Constantine's hand in the scene showing his victory, which is immediately to the right.

*Constantine's dream* is a naturalistic night scene. The foreground is seen by the light flowing from the cross held by the steeply foreshortened angel in the upper left part of the picture. Unfortunately, the cross itself having originally been shown in gold, its colour has now vanished and the small shape can only just be made out, pale yellow against the pink of the tent, directly above the angel's wrist. The light from the cross is slightly golden. It illuminates the open front of Constantine's tent and the figures of the sleeping Emperor, his two armed guards and the attendant sitting beside the bed. The light is clearly not real in the normal sense since the impassive guards and the attendant, staring out blankly, plainly cannot see it. In the background we have night. Moonlight shows up the shapes of the tents of Constantine's army but gives only a bluish hint of their colour. The first written description of how faint illumination affects our apprehension of colours dates from the nineteenth century—the phenomenon is called the Purkinje Effect—but Piero had obviously noticed it four centuries earlier.

In *Constantine's dream* as in the *Baptism*, it is mainly simple relations of size that are used to give a sense of space. On the other side of the window to the left of *Constantine's dream* is a scene of the *Annunciation* (Fig. 5.2), in which mathematical perspective is much more in evidence. The figures are set against architecture of a classical style that is at least partly inspired by the decoration of the interior of the Baptistery in Florence (in which pieces of ancient work were in fact re-used). The doors at the back are similar to the doors that Ghiberti was engaged in making for the Baptistery (see Fig. 1.15). Rather than the usual spray of lilies, the angel is carrying a palm frond, as a symbol of Christ's victory over death. There are several other examples of angels equipped with palm fronds from the Arezzo region at about this date (the 1450s). Piero's patrons probably made the choice for him. Like angels in other pictures by Piero, this one seems to be conceived as an adolescent, slightly under adult size, but it seems to me to be more than mere contrast that makes the Virgin look too large. She appears in fact to be too large for the architecture. There are not enough suitable lines for one to attempt to reconstruct the perspective of the *Annunciation*, but it is obvious that the Virgin is under the loggia, since part of her cloak is hidden by the column and part of the pavement is visible between her and us. It is possible that Piero's patrons simply insisted that the Virgin must be one of the largest figures in the whole cycle of frescos, but one wonders, in that case, why they had given the commission to an acknowledged expert in naturalistic perspective.

It was presumably as an acknowledged master of this aspect of his craft that Piero chose to write his treatise only about that particular part of painting. After the introduction to the first book, we hear no more about colour, and there are very few explicit references to *disegno* (a word whose meaning covers design and composition as well as the drawing of individual elements in the picture). The main text of the first book begins with Euclidean *perspectiva*, that is natural optics, the first proposition being that 'Every quantity presents itself to the eye as subtending an angle.' As Piero immediately says, this is obviously true, but there follow about a dozen lines of discussion and an appropriate diagram. The following six propositions all belong to standard Euclidean optics, and Piero gives references to the *Elements* (for instance to *Elements* Book 1, Proposition 24, from

**Fig. 5.2** Piero della Francesca, *The Annunciation*, from *The Story of the True Cross*, fresco, San Francesco, Arezzo.

Piero's Book 1, Proposition 3) and to one of Euclid's optical texts (from Piero's Book 1, Proposition 6).

Piero's first original theorem appears as Proposition 8. It is unspectacular, but as we shall see it turns out to be important. It is

If above a given straight line divided into several parts a line be drawn parallel to it and from the points dividing the first line there be drawn lines which end at one point [i.e. are concurrent], they will divide the parallel line in the same proportion as the given line.

A copy of Piero's diagram is shown in Fig. 5.3. We have put all the lettering into upper case since upper case letters are used in Piero's text. Piero's proof is in a style with which any reader of fifteenth-century mathematical texts rapidly becomes familiar: it relies upon similar triangles. After describing how to draw the diagram—in the manner of the *Elements* rather than the abacus book—Piero explains what he means by the parallel line *HI* being divided in the same proportion as *BC*.

I say it is divided in the same proportion as the given line .BC. because .BD. is to .DE. as .HK. is to .KL., and .EF. to .FG. is as .LM. to .MN., and .FG. to .GC. is as .MN. to .NI., ….

Then, following the comma, we have the proof

…, and the triangle .ABD. is similar to the triangle .AHK., as is .ADE. to the triangle .AKL., and .AEF. is similar to the triangle .ALM., so that they are proportional and the proportion that obtains between .AB. and .BC. is that of .AH. to .HI., the larger sides being proportional, the smaller sides are proportional and the angles of the triangle .ABD. are similar to [i.e. the same as] the angles of the triangle .AHK., so [the triangles] are proportional, as is shown in Euclid [*Elements*] Book 6, Proposition 21; and similarly for the others [i.e. other triangles], which is what is proposed.

Apprentice painters unfamiliar with *Elements* Book 6 (probably in a large majority) presumably had to take some of this on trust, but to the modern reader there is nothing very remarkable about this proof. What we need to have noticed, it will later turn out, is that the proof also works backwards, that is, the converse of the theorem is also true.

There follow three more theorems about establishing proportionalities. The proofs are all in a style similar to the above, but at the end of the last one Piero proposes a numerical example and then goes on to discuss further numerical examples subjected to 'degradation' by the laws of natural optics. The next section after that is the first to deal with artificial perspective.

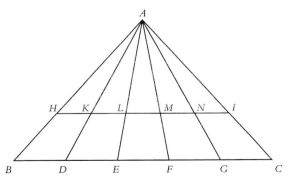

**Fig. 5.3** Copy of figure for Piero della Francesca, *De prospectiva pingendi*, Book 1, Proposition 8, with lettering in upper case.

# Perspective of plane figures

Piero's first theorem concerns an area, which later propositions develop into an Albertian *pavimento*. Piero's approach is, of course, quite different from Alberti's, since he is in principle addressing himself to practitioners. His proposition is headed

From the given eye [position] and on the determined plane to degrade the assigned surface.

Translation of the heading is awkward because Piero is using technical terms, explained in his introduction, whose modern equivalents are constructed rather differently. For instance, my 'determined plane' translates '*termine posto*', Piero having defined '*termine*' (whose meanings include limit and boundary) in such a way as to make it clear that he means the picture plane. There was no standard vocabulary for this kind of thing in Piero's time, or for a long time thereafter. Authors who mean to be helpful, as Piero does, take the precaution of explaining what their terms mean. One piece of his vocabulary is, however, of wider significance: the use of the verb 'to degrade' (*degradare*) to signify the changes brought about by showing a shape in perspective. The original shape is referred to as in its 'proper form' (*propria forma*) or as 'perfect' (*perfetto*), and the perspective version is called 'degraded' (*degradato*). This usage—though sometimes with slightly different wording—is common to all subsequent treatises on perspective. It is of interest because it shows where attention is being directed, namely to the change brought about by perspective. Moreover, degradation has a sound of being a one-way process. To the twentieth-century eye it is clear that the procedure of perspective involves a form of conical projection in which the vertex is the eye. The Renaissance view of the matter is less abstract: attention is focused on the specific practical problem, that of drawing the changed shape.

The diagram for this first perspective theorem, Book 1, Proposition 12, is as shown in Fig. 5.4. The lettering is given by Piero's drawing instructions. The point *A* is the position of the eye, *BC* the 'assigned surface' (*piano assignato*), *B* the *termine posto*, that is, in this case, the point in which the picture plane intersects the line *DC*. Piero asserts, and then proves, that *BE* is the 'degraded surface' required, since it subtends at *A* an angle equal to that subtended by the 'assigned surface' *BC*. He does not explain his diagram as showing a vertical section. Very similar diagrams have been used to illustrate the earlier propositions derived directly from Euclidean optics, so this one presumably seems self-explanatory. One might compare it with the Euclidean set-up shown in our Fig. 1.2. A diagram very like this one has appeared in the earlier part of Piero's book. Of course, using a line to represent the surface also has the advantage of allowing Piero to leave the shape of the surface somewhat indefinite.

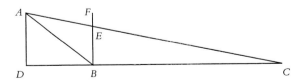

**Fig. 5.4** Copy of diagram for Piero della Francesca, *De prospectiva pingendi*, Book 1, Proposition 12.

A surface of definite shape appears in his next proposition:

To make the degraded surface a square.

This time the drawing instructions completely dominate the proceedings, and result in a diagram like that shown in Fig. 5.5, in which we have adopted Piero's lettering completely, thereby obtaining two points lettered *A*, two points lettered *D*, and two points lettered *E*. The first drawing instructions of the proposition have established that, as in the previous diagram, *AD* and *BF* are perpendicular to the line *DBC*. The further drawing instructions make *BCGF* square, *AI* the perpendicular bisector of *BC* and the lines *AA* and *EDEK* parallel to *DC*. The versions of this diagram that appear in different manuscripts of Piero's treatise are not all exactly the same, but all of them appear to omit some of the lines which we have been instructed to draw. Some of the apparent omissions may be due to the fact that photographs do not show up lines scribed into the paper rather than drawn on it, so scribed construction lines may disappear in the versions of diagrams published by modern scholars. In any case, some of the points shown in Fig. 5.5 do not play any part in the reasoning of Piero's proof, so we may, for clarity, work from the simplified version shown in Fig. 5.6. The figure bears a fairly strong resemblance to that for the Albertian construc-

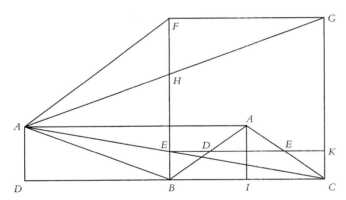

**Fig. 5.5** Diagram for Piero della Francesca, *De prospectiva pingendi*, Book 1, Proposition 13, obtained by following Piero's drawing instructions. There are two points each with the letters *A*, *D* and *E*.

tion (Fig. 2.4), though Piero's construction has only given us one transversal, namely the line *DE*, which represents the back edge of the square. To facilitate comparison a corresponding diagram for the Albertian construction is shown in Fig. 5.7. As we saw in Chapter 2, it is fairly easy for a twentieth-century mathematician to conceptualise the Albertian diagram as being a combination of a section and the picture plane. Piero's drawing instructions imply an underlying conceptualisation in rather different terms, and his proof that the construction is correct in fact merely uses the plane diagram without reference to the original spatial significance of its components. As we shall see, although Piero's diagram (unlike Alberti's) contains some elements of a ground plan, we have to make some quite complicated assumptions in extracting them.

The first part of the diagram to emerge from Piero's drawing instructions is shown in Fig. 5.8. Here *BCGF* is the square area that is to be represented in degraded form, and *A* is the eye. So if

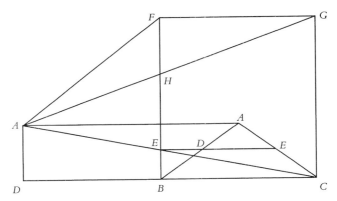

**Fig. 5.6** Diagram for Piero della Francesca, *De prospectiva pingendi*. Book 1, Proposition 13, simplified for clarity.

the picture plane is at *BF* the back of the square *GC* will be seen as *HE*, because these lengths subtend the same angle at *A*. That is what Piero says. He is clearly assuming that we know that the result still holds if *A* is not in the ground plane *BCGF* but some way above it, for the actual three-dimensional situation, of which Piero does not provide an illustration, is as shown in Fig. 5.9. Moreover, the part of Piero's diagram Fig. 5.5 that looks as if it were a view of the picture plane,

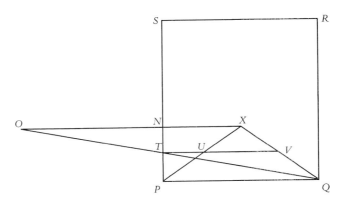

**Fig. 5.7** Albertian construction for the back edge of a square of side *PQ*. Compare with Fig. 2.4, p. 26.

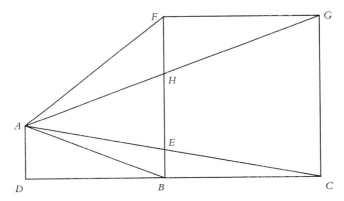

**Fig. 5.8** The first part of the diagram for Piero della Francesca, *De prospectiva pingendi*, Book 1, Proposition 13.

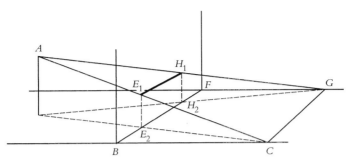

**Fig. 5.9** Three-dimensional diagram for the problem of drawing a square *BCGF* in perspective (*De prospectiva pingendi*, Book 1, Proposition 13), with the eye at *A* and the picture plane a vertical plane through *BF*. There is no corresponding diagram in manuscripts of Piero's treatise.

namely the isosceles triangle *ABC*, implies that the position of the eye is opposite the centre of the line *BF*, whereas the quasi-ground plan, shown separately in Fig. 5.8, implies that *A* is not central. Presumably Piero expects us to know that movement of *A* along a line parallel to *BF* will not affect the length *EH*. We may note also that, although Piero's drawing instructions began by telling us to draw the figure as for the previous diagram, the point *A* now no longer lies in the vertical plane through the line *BC*. All in all, Piero is asking quite a lot of his readers. Some of it may reflect drawing conventions with which fifteenth-century readers could be expected to be familiar. In any case, it encourages the twentieth-century reader to take a suitably wary attitude to the remainder of Piero's proof.

Since *EH* has been shown to be the appropriate length for the representation of the furthest edge of the square, what Piero needs to show is that the segment *DE* forming the upper edge of the trapezium in the isosceles triangle *ABC* is equal to *EH*. This is done by means of similar triangles. The pair of similar triangles *AAC*, *EEC* is used as a link between the pairs *AGC*, *AHE* and *ABC*, *ADE*. The actual proof is very short, merely asserting the equalities of certain ratios. Here, Piero's double use of letters makes some of what he says ambiguous, but closer inspection shows that, when this is so, both versions are in fact correct. The double lettering has made the proof shorter. This may have been regarded as an elegance by fifteenth-century mathematicians, but it has proved to be a barrier to understanding in the twentieth century (even leading one art historian to assert that Piero's proof is incorrect). The passage hardly seems to have been designed to be understood by an apprentice painter.

Once we know that the upper edge of the trapezium is the right length to represent the back edge of the square, symmetry ensures that the other two edges are as shown by the appropriate segments of the lines joining *A* and *B* to *C*. It is clear that this point *A* corresponds to the centric point *X* in the Albertian diagram shown in Fig. 5.7, but Piero does not call it a centric point. Perhaps he felt this name would be taken to imply that *A* must be in a central position, which, in the final lines of his proposition, he points out need not be so, though he adds that *A* must be within the picture (which need not be so either).

In the proposition we have been discussing, the point *A* inside the square was constructed simply as lying on the perpendicular bisector of *BC*, at a height equal to the height of the external point *A* above the line *CB* produced. That is, the point was not constructed by reference to ortho-gonals (as Alberti had, apparently, constructed the corresponding point in his system). Piero's next

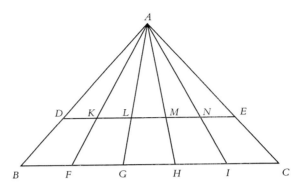

**Fig. 5.10** Copy of diagram for Piero della Francesca, *De prospectiva pingendi*, Book 1, Proposition 14, showing division of the square by orthogonals.

proposition, Book 1, Proposition 14, introduces orthogonals as lines dividing the square into equal parts. We begin from the degraded square, lettered as in the previous proposition, with the eye described as being at *A*:

As it was said, let *.BCDE.* be the square and the eye be *.A..* Divide *.BC.* into as many parts as you like it divided, in *.FGHI.* equally [say], then join *.F.* to the point *.A.* and *.G.* and *.H.* and *.I.* to the point *.A.*, which lines will divide *.DE.* in the points *.KLMN..* I say that *.DE.* is divided in the same proportion as that in which *.BC.* is divided, because … .

Piero then repeats the explanation of proportional division that we encountered among his mathematical preliminaries (see above). The diagram we should obtain by following these drawing instructions is shown in Fig. 5.10. We are now working entirely in the picture plane, and the construction can be carried out in the actual picture field—Piero having specified that *A* should lie within it. The fact that the division along *DE* is in the same proportion as that along *BC* has been established by the preliminary theorem that we discussed earlier. In terms of construction procedures, that is all we need to know, but Piero's having gone to the trouble of proving this apparently rather trivial theorem suggests that he may have recognised the importance of its converse, namely that all orthogonals converge to *A*.

The next proposition, number 15, completes the *pavimento* by putting in the transversals. Piero begins, as usual, by delivering a long series of drawing instructions, which tell one to draw the diagonal *BE* and through the points in which it cuts the transversal draw lines parallel to *BC*. As proof of the correctness of this, Piero then instructs one to draw the square 'in its proper shape' (*in propria forma*) below the degraded one, and carry out the analogous process in this new square. The result is a diagram like that shown in Fig. 5.11. This time, the double lettering seems quite reasonable. It neatly shows up the fact that in order to join the two parts of the diagram along *BC* Piero has produced a kind of mirror image. Perhaps to him it would have looked as if the lower part had been folded forward along *BC*, having originally been in a plane perpendicular to its present one.

If we compare Piero's method of constructing the *pavimento* with the Albertian method and the distance point method, we see that it is not exactly the same as either of them. Like Alberti, Piero requires the distance between the eye and the picture to be represented by a length measured beyond the edge of the picture field. However, this point is only used once, to construct the fur-

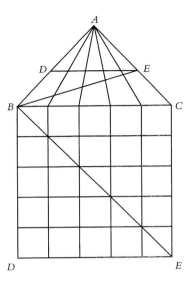

**Fig. 5.11** Copy of diagram for Piero della Francesca, *De prospectiva pingendi*, Book 1, Proposition 15, showing the procedure for completing the *pavimento*.

thest transversal. After that, Piero uses the diagonal, which Alberti proposed as a check on transversals, to construct the remaining transversals in his diagram. Accordingly, after the initial transversal has been constructed, giving the degraded shape of the square, the remainder of the construction can be carried out within the picture field. This avoids the awkwardness that would be introduced by the errors that are unavoidable if a small intricate picture of a *pavimento* had to be transferred to a panel by means of a preparatory drawing. A good example of such a *pavimento*, apparently drawn with great accuracy, can be seen in Piero's *The Flagellation of Christ* (see Fig. 5.20, p. 101), which will be discussed below.

After the basic square-tiled *pavimento*, Piero goes on to deal with the figures produced by making various forms of uneven division along *BC*, and then to consider polygons within the square. In each case, the 'proper shape' of the figure is shown below the 'degraded' one, as in our Fig. 5.11. Readers presumably found this helpful, since the same device is adopted in many later treatises on perspective. The example of a regular octagon (in fact the plan of an octagonal building, so including the thickness of the wall), Piero's Proposition 29, is shown in Fig. 5.12. Piero has treated the most general case, in which none of the sides of the polygon is parallel to a side of the square *BCED*. Similarly general cases are treated for other polygons. Piero has, however, already treated one special case for the octagon: in Proposition 16 he considered the regular octagon formed by cutting off the corners of the square *BCED* (and without explaining how to ensure that the octagon obtained in this way was regular). It is tempting, but probably rash, to see this proposition as Piero's reconstruction of how Brunelleschi may have drawn the ground plan in his panel showing the Baptistery. Of course, Piero may have actually seen Brunelleschi's panel during one of his visits to Florence, some of which were made in Brunelleschi's lifetime. There is no evidence the two men did ever meet one another, but Brunelleschi's panels were for some time in the possession of the Medici family in Florence, where Piero might have had access to them.

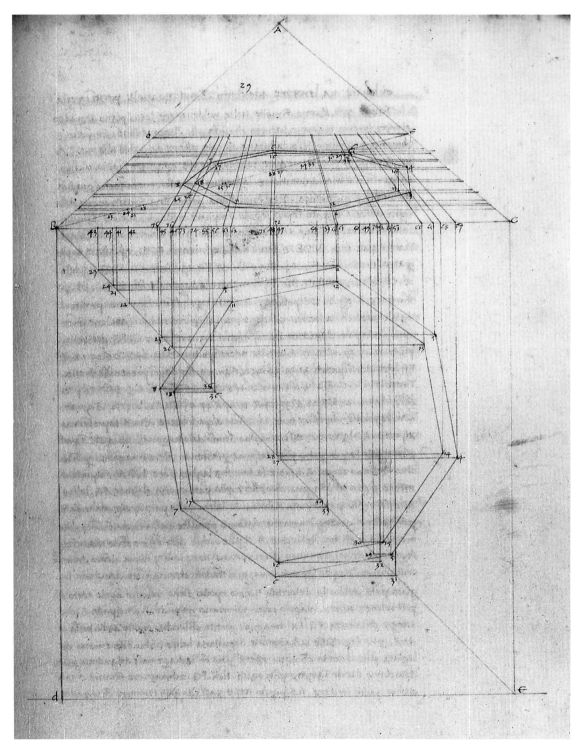

**Fig. 5.12** Diagram for Piero della Francesca, *De prospectiva pingendi*, Book 1, Proposition 29, drawing the plan of an octagonal building in perspective. From the manuscript in Parma (Biblioteca Palatina, no. 1576).

The final proposition in the first book of Piero's treatise, Proposition 30, is not a drawing problem but a theorem. Moreover, it is a theorem Piero appears to take rather seriously. It is introduced as follows:

To remove the error made by some who are not very experienced in this science (*scienza*), who say that often when they divide the degraded surface into units (*bracci*), the foreshortened one comes out longer than the one that has not been foreshortened; ... .

The Tuscan word '*scurto*', which I have translated 'foreshortened', has an etymological connection with the word '*curto*', meaning 'short', hence the paradoxical appearance of a part subjected to this process coming out longer instead of shorter. Luckily, the corresponding English words convey the same impression. Mathematically, however, the paradox is meaningless. There is simply no reason why the imposition of perspective should not make things appear longer than they really are. In fact, it sometimes does, as can be seen in Fig. 5.13, in which Piero's method of construction has been used to obtain the first transversal, *ST*. It is clear that the 'degraded' *BS* in the upper part of the figure is longer than *BS* in its 'proper form' in the lower part. So, mathematically speaking, Piero would appear to be backing a loser. His meaning becomes clearer if we read further, for he adds:

and this happens by not understanding the distance there should be from the eye to the limit (*termine*) where the things are put [i.e. the picture plane], nor how wide the eye can spread the angle of its rays; so they [i.e. the inexperienced] suspect perspective is not a true science (*vera scientia*), judging falsely because of ignorance.

That is, Piero claims that this effect of something coming out longer is due to the angle the picture is made to subtend at the eye, whose 'rays', that is the beams the eye emits in order to see, cannot embrace more than a certain angle. The angle subtended by the picture becomes large when the

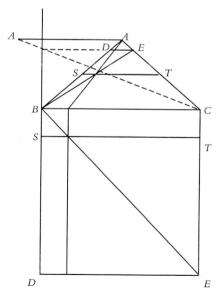

**Fig. 5.13** Diagram to show 'distortion', in Piero's sense. The length of *BS* shown in perspective in the upper part of the diagram is longer than the true length of *BS* shown in the lower part.

viewing distance becomes small—which was the case in Fig. 5.13, where the viewing distance, the distance between the point $A$ on the left of the diagram and the vertical line through $B$, has been made very short. The reference to the actual working of the eye is a reminder that to Piero his perspective for painting is an extension of ordinary *perspectiva*, whose business included the whole science of sight, the human factor not excepted. As we shall see, this appeal to what we should now call physiological optics plays a significant part in Piero's proof that perspective genuinely is a 'true science'. Indeed, in order to see the point of Piero's proof at all, one needs to know that the maximum angle the (immobile) eye was believed to be able to embrace with its rays was a right angle. Perspective is, as it were, giving us the geometry of a single glance.

This seems extremely artificial to a twentieth-century reader. We know the eye does not stay still while looking at a picture, and that the width of the field of view is considerably greater than a right angle. It is sometimes hard to believe that these things were not obvious to fifteenth-century people also. However, there is overwhelming literary evidence—that is, the whole contemporary literature on the relevant parts of natural philosophy—to show that they were not. One may be permitted to wonder why Piero, as a mathematician, should have worried about the verbal rather than mathematical paradox of the foreshortened length coming out longer, but he is entirely conventional in accepting that perspective deals with a single glance and that the field of view of the eye is a right angle. Moreover, it was at the time universally accepted that it was enough to consider one eye. The two eyes were believed to act exactly as one, combining their information before the mind made use of it. (Usually the combining process was believed to take place at the optical chiasmus, the part of the brain where the nerves from the eyes cross.) Since it was generally agreed that nature did nothing in vain, our having two eyes was explained as Divine Providence guarding against the loss of sight by supplying an extra part.

Thus for Piero the worries of the 'inexperienced' cannot merely be dismissed. He accepts the criteria they imply and sets about proving that perspective is a true science. That is, he sets out to prove that if the viewing distance is chosen so that the picture subtends a right angle, or less, at the eye, then a foreshortened orthogonal line segment cannot come out longer in the picture than it is in reality.

As usual, Piero begins by giving detailed instructions for drawing the diagram. Its finished form is shown in Fig. 5.14, in which the eye is at $A$, in the centre of the array of squares. In this diagram, Piero has used numbers rather than letters for some of his points. In fact, this has been happening from Proposition 17 onwards. We now begin to see the usefulness of his habit of putting a dot before and after the names of points, lines and areas in his text. The part of the diagram that we need for the first part of Piero's proof is the upper left corner, which has been redrawn in Fig. 5.15. The square *.BKL21.* has been constructed to be the same size as all the other small squares surrounding the large one, and *.AK.* has been drawn to cut *.F21.* in $M$. The position of the picture plane is given by the line $FG$, but the picture is now imagined as extended so as to include $L$. What we have is a quasi-ground plan, as for Proposition 13, in which we are assumed to accept that nothing important is changed if the sight lines are imagined as coming from a point vertically below the eye.

Piero's proof is disconcertingly short. He merely says that as seen from $A$, the point $K$ will be seen at $M$, so the length $KL$ will be represented in the picture plane by $LM$, which is larger than *.L21.*, so the foreshortened length has come out longer, but in fact the eye at $A$ cannot see $K$ because the limit of its vision is defined by the diagonal $AB$, so all is well. Unfortunately, what

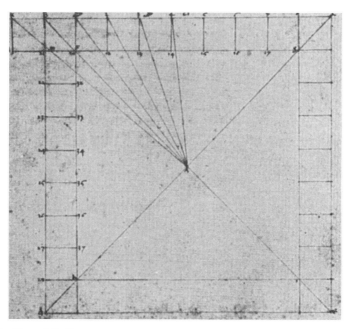

**Fig. 5.14** Diagram for Piero della Francesca, *De prospectiva pingendi*, Book 1, Proposition 30. From the manuscript in Parma (Biblioteca Palatina, no. 1576).

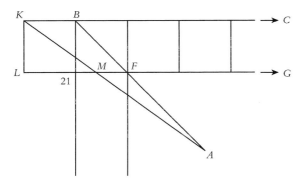

**Fig. 5.15** Copy of upper left part of diagram for Piero della Francesca, *De prospectiva pingendi*, Book 1, Proposition 30.

Piero has proved, though true, is not what he needs to prove. He has proved that if $K$ lies beyond the diagonal $AFB$ then the length of the segment of the orthogonal through $K$ cut off by the first transversal will be greater than $KB$. What he actually needs to prove is that this is so *only if* $K$ lies beyond the diagonal.

It turns out that this result is not true. At least, it is not true unless the eye is truly imagined as being in the ground plane of the picture—which clearly entirely disposes of any problem of constructing a *pavimento*. To see this, I needed to go back to a diagram like Fig. 5.13 and do a certain amount of algebra on the geometrical problem it presents. Until the very last, when I used the tangent function of my calculator to turn a minimum viewing distance into a maximum viewing angle, all that was required was Pythagoras' Theorem and a knowledge of the proportionalities obtainable from similar triangles. In Piero's time, circular functions were used by astronomers (though not quite in the form in which we use them today) but are not found in abacus books (as far as I know). Indeed, the very last problem in Piero's book on the five regular solids—the one about a circle and the three sides of a triangle that are all tangents to it (see above)—involves him in an elaborate series of calculations which could have been considerably shortened if he had had recourse to the sine rule, and a table of sines. So perhaps my use of the tangent to find the angle was anachronistic. However, the rest of my work was straight out of the abacus school.

Historians of science get used to meeting a nice class of intellect among the illustrious dead, and one of the rules of the game is accordingly never to be hasty in coming to the conclusion that the person one is reading is more of a fool than oneself. The proviso is that one has to keep a weather eye open in the cases where what has apparently been proved fits in very well with an established system of some kind. The acceptance of Piltdown Man is a good example from our own century. The system into which Piero is fitting his 'perspective for painting' is *perspectiva* proper, the complete science of vision, and the purpose of the reasoning in Book 1, Section 30, is clearly to show that the new geometrical technique is a valid extension of this established science.

Drawing a few diagrams shows that Piero's 'theorem' is very nearly true, for eye heights like those used in the earlier part of Book 1, and that the rule he actually recommends, namely that the picture should never subtend more than 60° at the eye, will work for all the eye heights he uses in his problems. The discordance between what Piero claims to have proved and the rule he recommends in practice (without further explanation) suggests that he may have realised the proof was faulty. Why should he then not have provided a rigorous treatment of the problem? After all, as we have seen in the proof of the correctness of the construction for the image of a square (Book 1, Section 13), when the result was sufficiently important he seems to have been prepared to present material aimed at fellow mathematicians rather than the apprentice painter to whom the text as a whole appears to be addressed. Perhaps the problem in Section 30 really did seem too difficult. As we have seen, its diagram represents a breakdown of the usual condition that a quasi-ground plan will do instead of a truly three-dimensional diagram (a condition that was met in Section 13). Moreover, though soluble by abacus book methods, if we decide to consider viewing distance and establish limits for the maximum viewing angle rather than calculating its exact size, the problem is considerably more complicated, in mathematical terms, than anything else in Piero's treatise. So maybe he decided something shorter and simpler was more appropriate, even if not quite true. History provides an excellent Machiavellian defence: Piero's proof was printed, almost word for word, in the next century, in a work that was widely read—Daniele Barbaro's *La pratica della perspettiva* (Venice, 1568, 1569), which we shall discuss in the next chapter—and for a little over four centuries thereafter no-one noticed that anything was wrong with it, or at least no-one said so in print. If Piero made that kind of calculation, namely that the proof would pass muster with his intended readership, then he got it right.

Piero returns to the problem of designing a wide picture at the end of his second book (Book 2, Proposition 12). This time, he is concerned with a row of columns running across the picture. Will the ones at the edge come out wider than they are in reality? Presumably the problem found its way into the second book because Piero is now dealing with architecture. The problem of the colonnade, as treated, is none the less two dimensional, and therefore properly belongs with the

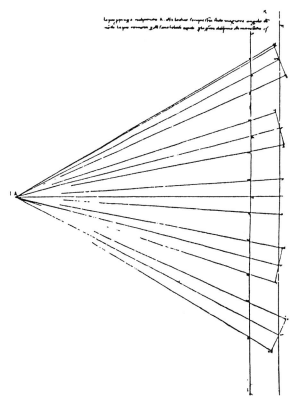

**Fig. 5.16** Diagram for Piero della Francesca, *De prospectiva pingendi*, Book 2, Proposition 12, showing a row of columns. From the manuscript in the British Library (Add. Ms 10366).

work of Book 1. Piero's conclusion is that the columns at the edge will not come out too wide. One version of the corresponding diagram is shown in Fig. 5.16. The proof is approximate, though Piero does not say so. However, in today's terms, the approximation consists merely in leaving out some inconvenient sines and cosines. The theorem itself is true, as far as I can see.

## Prisms and combinations of prisms

The second book of Piero's perspective treatise essentially deals with prisms and combinations of prisms, though some of the shapes are elaborated into forms useful in the practical business of painting.

The introduction to the book is only a few lines long. It begins

A body has three dimensions: length, width and height; its limits (*termini*) are surfaces.

We then have a list of possible shapes, which is helpful to a modern reader as introducing Piero's technical terms. Again we are dealing with vocabulary that was not entirely standardised, so the introduction may have served a similar purpose for its original readers. In fact, what Piero says here is almost identical with his introduction to the section on three-dimensional geometry in his abacus treatise.

The first body to be considered is a cube with one edge parallel to the ground line of the picture (*BC* in Piero's earlier diagrams). There at once follows a cube in a more general orientation, having none of its edges parallel to the ground line. Higher polygons are then used as ground plans, including one with so many sides (sixteen) that it makes an appropriate shape for a column, as shown in Fig. 5.17. This is followed by three hexagonal prisms superimposed to make a well head surrounded by steps (see Fig. 5.18).

The drawing instructions for these and the following propositions omit not a single detail. Piero seems not to know the words 'and so on', but simply gives instructions for drawing every line. The monotony of this is mind-numbing for a mere reader, but perhaps the repetitive style might have been helpful to an apprentice who was actually carrying out the instructions as he went along. Visualising forms in space with this degree of mathematical precision is certainly not as easy for everyone as the illustrations to Piero's treatise make it look.

The well head, a fairly realistic example, is followed by a cube that has had mouldings added at top and bottom (and looks rather like a pagan altar), then by an octagonal prism placed with its

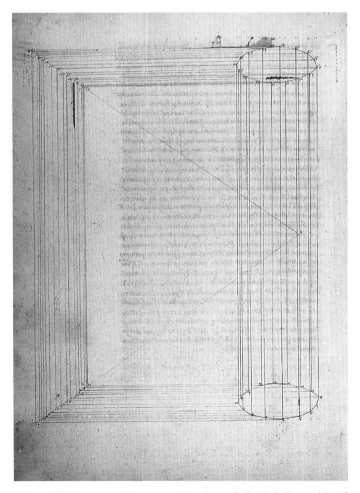

**Fig. 5.17**  Diagram for Piero della Francesca, *De prospectiva pingendi*, Book 2, Proposition 5, showing a column. From the manuscript in the British Library (Add. Ms 10366).

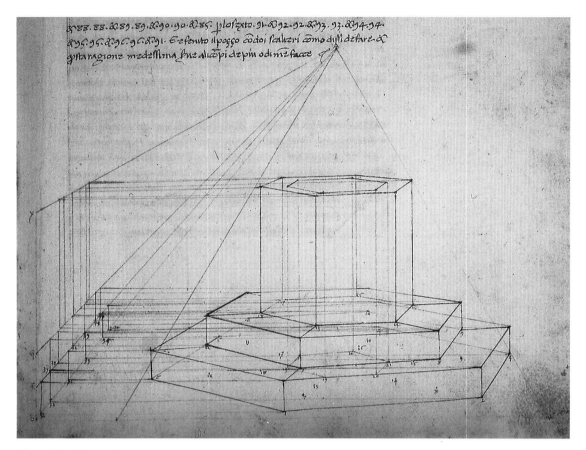

**Fig. 5.18** Diagram for Piero della Francesca, *De prospectiva pingendi*, Book 2, Proposition 6, showing hexagonal prisms superimposed to make a well head on a plinth of steps.

very long principal axis horizontally but at an angle to the picture plane, and then by a cube that has become a house, but still retains the construction lines that belong to its basic geometrical shape (see Fig. 5.19). We have a more or less standard fifteenth-century town house, such as are shown in many paintings, including some by Piero himself (see Figs 5.20 and 5.26). After the house comes 'a temple with eight faces' (none of them parallel to the picture plane, since the base is the octagon from Book 1, Proposition 29, shown in our Fig. 5.12 above). This is followed by a square structure whose roof is in the form of a cross vault. The book ends with the proposition about the row of columns which was mentioned above in connection with Piero's discussion of the permissible width for a picture.

Mathematically, Piero's second book is much less interesting than his first, but its examples seem to have been very well chosen, for almost all of them are to be found repeated, sometimes in simplified forms, in the numerous perspective treatises addressed to artists that were printed in the sixteenth and early seventeenth centuries. A fair number of the elements Piero has described in his first two books can be found in his small panel picture *The Flagellation of Christ*, shown in Fig. 5.20. The perspective scheme of this picture not only looks accurate but has been proved to be so. Detailed measurements have allowed a reconstruction of the ground plan of the scene and any

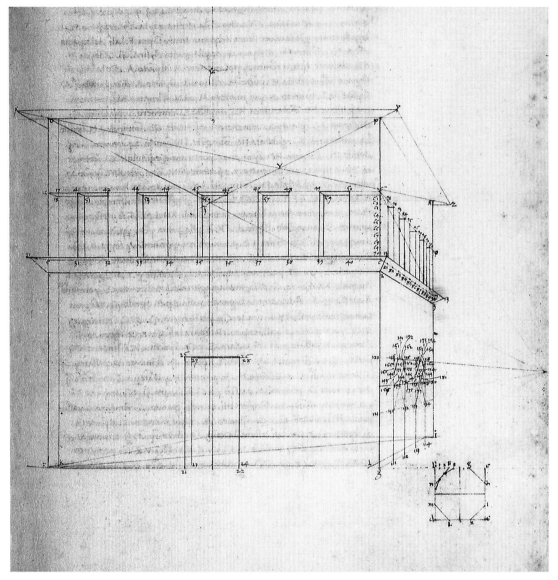

**Fig. 5.19**  Diagram for Piero della Francesca, *De prospectiva pingendi*, Book 2, Proposition 9, showing a cube elaborated to become a house. From the manuscript in the British Library (Add. Ms 10366).

number of sections. Of course, a few elements elude reconstruction—for instance, the lower edges of the buildings on the right cannot be seen, so their exact positions cannot be found—but the total effect is of extreme precision, and of real pictorial space. The only departure from mathematical correctness seems to be that the strip of white marble that runs into the picture under the bases of the row of columns (and, with the nearest column, serves to divide the part of the picture plane containing the background scene from that containing the foreground one) has been made slightly

**Fig. 5.20**   Piero della Francesca, *The Flagellation of Christ*, tempera on panel, 59 × 81.5 cm, Galleria Nazionale delle Marche, Urbino.

narrower than mathematics would dictate. This departure from correctness is presumably deliberate: Piero no doubt preferred the band of white to be a little narrower, and guessed the change would not be noticeable to an ordinary viewer.

In *The Flagellation of Christ*, element after element can be matched with corresponding propositions in Piero's treatise on perspective: we have two patterned pavements, one simple, one extremely elaborate, plus a further quasi-*pavimento* provided by the beams and coffering of the ceiling of the Judgement Hall, as well as the modified prisms of the houses on the right, of the fluted shafts of the columns and of the steps of Pilate's throne (on the base of which Piero has signed his name). All the same, the picture contains a curious contradiction, which is that, despite its having a viewing distance of about two and a half times the width of the picture, the detail of the execution is such that no-one would wish to maintain that distance. Presumably because of the detailed illusionistic treatment, in such things as the reflection of light from the grey silk robe of the man with his back to us, going too close to the picture does not seem to have the customary destructive effect on the pictorial space. On one level *The Flagellation of Christ* can be made to seem an example of the use of mathematics in art, but on another it tells one at least equally clearly that Piero, like Masaccio before him, knew the difference between a theorem and a picture.

## More difficult shapes

Piero himself describes the third book of his perspective treatise as dealing with bodies whose complicated shapes make them 'more difficult'. These bodies turn out to include the moulded bases of columns, their decorative capitals, and human heads. Because the shapes are more difficult, Piero says, he will use a different method. This method is, effectively, ray tracing—and its application is so laborious that the monotony of the drawing instructions in Piero's second book pales in comparison. The drawing instructions in this third book seem ready-made for the electronic computers that will be invented nearly five centuries later.

Perhaps because he is aware that this part of his work is not following established procedures, Piero provides his third book with a relatively long introduction, taking up more than a page. In fact, the introduction to the third book is considerably longer than that to the first one. In it, Piero not only explains that perspective is a way of attaining to visual truth but also, in the best humanist manner, asserts that it was crucial to the wonderful works of art produced in ancient times. He naturally lists the famous painters of those times, most of the names apparently being derived from Vitruvius' references to painting in *On architecture*, though it is, of course, possible that Piero came across them in an intermediate source.

The first proposition of the third book of Piero's perspective treatise is extremely simple: to draw a square in perspective. The reason for this choice is that Piero considers this is a good way of introducing his method. As he puts it

Now, to demonstrate the method which I intend to follow, I shall give two or three demonstrations for plane surfaces, so that through them we can arrive more easily at finding out the degradation of bodies.

This seems to be a sound pedagogical procedure and should presumably be taken to indicate that Piero does actually regard this different method as a practical one—however unwieldy it may seem to us.

The first step is to draw a diagram of the square in its 'proper form', *BCDE*, then choose point *A*, 'which will be the eye', placing it at the required viewing distance. The next instruction is

In the point .*A*. is fixed the nail, or if you want a needle with a very fine silk thread, it would be good to have a hair from the tail of a horse, particularly where it has to rest against the ruler; … .

This is the first we have heard of nails and so on. As one reads on, it turns out that the drawings at the end of the proposition, which have been copied in Fig. 5.21, do not represent merely drawings but a mixture of drawings and other apparatus. The squares shown on the right in the first two parts of the figure are indeed drawings. The upper one shows the square edge on, but for the sake of clarity the actual shape has been shown as well, as if folded upwards into the plane of the diagram (this is a quite normal device in fifteenth-century technical drawings). The second drawing shows a ground plan. In both these first two parts of Piero's figure, the lines to the left of these drawings of the square represent the picture plane. It is called *FG*, so Piero is not attempting to make the lettering in this proposition correspond with that used in the application of his previous method to the square in Book 1, Proposition 13 (see Figs 5.5 and 5.6). Moreover, in the example in Book 3, the square is placed a little back from the picture plane, though with one side parallel to the ground line, whereas in Book 1 one side of the square, the side *BC*, lay along the ground line. Further, in Book 3, the point in the picture plane directly opposite the point on the ground vertically below the eye is included in the diagrams, and given a name, *M*.

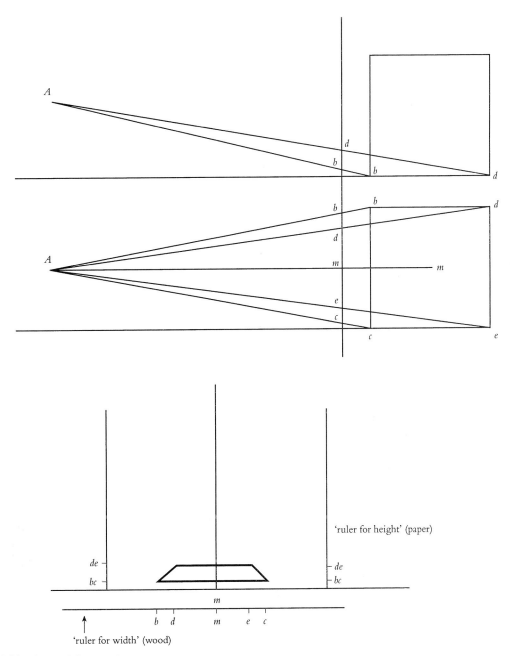

**Fig. 5.21** Copy of diagram for Piero della Francesca, *De prospectiva pingendi*, Book 3, Proposition 1, showing Piero's second method used for drawing a square in perspective.

The line representing the picture plane is part of the drawings, but is used only as a means of positioning what Piero calls 'rulers' (*righe*). These are flat strips of paper or thin wood that are used to record the positions in which the silk thread or horsehair attached to the nail or needle at *A* and positioned so as to pass through the points *B*, *C*, *D*, *E*, or *M* cuts the line representing the picture plane. We are recommended (in the imperative) to use a wooden ruler to record the widths, that is the positions corresponding to *B*, *C*, *D*, *E* and *M* found from the ground plan, and a paper ruler to record the heights, that is the positions found from the section.

To judge from the third part of Piero's figure, we have in fact made two paper rulers for the heights, since one is shown on each side of the finished drawing. Below, we have the wooden ruler for the width. Clearly, the rulers have been used to provide measurements which define the positions of the points *B*, *C*, *D*, *E* in the finished drawing. The point *M* has been used to line things up correctly. Piero describes the process in detail in his text (nearly two pages of it). To the twentieth-century eye, he seems to be using something like a rectangular coordinate system. This thought is not entirely anachronistic: Piero presumably knew about systems of latitude and longitude used to give positions of stars and planets in relation to the ecliptic in astronomical tables, or the positions of cities in relation to the terrestrial equator in the lists supplied in books on geography. Terrestrial longitudes were very difficult to determine at all accurately. Some of the values given look like guesswork. However, serious books on geography, including the famous *Geographia* of the astronomer Claudius Ptolemy (*fl.* AD 129–141), rediscovered in the early fifteenth century, doggedly supply rubbishy longitudes together with relatively good values for latitudes calculated from the height of the Sun at noon. Living where and when he did, Piero would have had plenty of opportunities of coming across this kind of coordinate system.

After dealing with the square, Piero next considers the regular octagon with one side parallel to the ground line but, like the square, set back a little from it. As in the corresponding proposition in Book 1, he assumes 'tu' already knows how to turn a square into a regular octagon by cutting off its corners. (The method given in later 'practical geometry' texts is to take half the length of the diagonal of the square, and taking each of the vertices of the square as centre mark this length off on each of the sides. The eight points so obtained are the vertices of the required regular octagon.)

After the octagon, Piero considers four concentric circles. At least, that is what his first diagram shows. Each circle is divided into twelve equal parts (an easy task) and the perspective construction is carried out for the resulting four concentric regular dodecagons. 'Tu' is presumably expected to put in the degraded forms of the circles by drawing a smooth curve through each set of twelve vertices. A similar method is used in dealing with all other curved lines that Piero encounters. Since 'tu' may be assumed to be a competent draughtsman, the method is an entirely reasonable one. Piero's use of it does not necessarily imply that he did not know that the degraded circles were ellipses. To draw an ellipse he would have needed to know, say, its two axes, and then to set up an appropriate drawing instrument. Ellipse-drawing instruments very probably were well known among mathematicians at the time, since they are described in a number of texts which cannot reasonably be suspected of displaying mathematical originality. However, even in the 1570s, Guidobaldo del Monte (1545–1607), a competent though generally unimaginative mathematician, thought it necessary to give an elaborate description of how to draw an ellipse (in connection with constructing some of the lines for a particular kind of astrolabe, which was a task for astronomers). So Piero's merely mentioning that his degraded circles were ellipses would not have been helpful to his putative readers, and explaining enough to make the knowledge useful would have made a very long digression.

After the circles, we move on to solid bodies. They include a cube in a general orientation (Proposition 5), a column base with mouldings (Proposition 6), and next a column capital like those shown in the Annunciation scene (Fig. 5.2) and *The Flagellation of Christ* (Fig. 5.20), in which we of course also find the moulded column base of Piero's previous proposition. Figure 5.22 shows

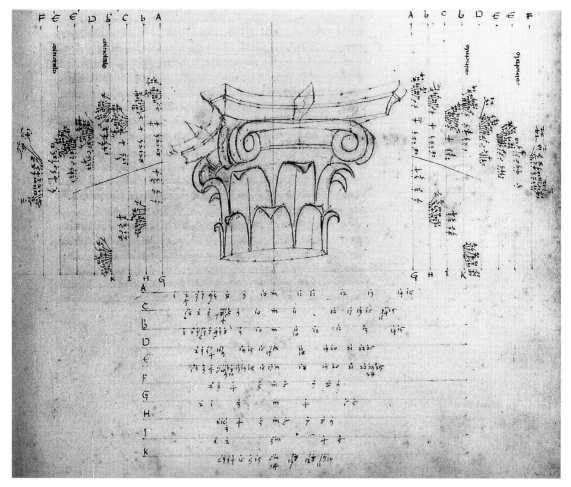

**Fig. 5.22** Diagram for Piero della Francesca, *De prospectiva pingendi*, Book 3, Proposition 7, showing the final drawing in the series for constructing the image of a column capital. Piero has used eight 'rulers for the height' and ten 'rulers for the width'. From the manuscript in Parma (Biblioteca Palatina, no. 1576).

the last in Piero's series of drawings for the column capital, the finished drawing with two sets of eight paper 'rulers for the height' to left and right of it and ten wooden 'rulers for the width' underneath it. The rulers have been lettered to show which goes with which. The points, other than the central one (*m*), have been numbered.

Next comes the human head. After that, the remaining three examples of the treatise deal with a coffered half dome, and two trick pictures. The first is a goblet that appears to stand up from the table on which it is painted, and the second a ring (the kind used to suspend lamps) that appears to hang down from a vault. Vasari, in his *Life* of Piero, says Piero did paint a *trompe l'œil* goblet on a table, but gives no details.

The drawings of the human heads are interesting because comparisons can be made with surviving works. The initial sets of drawings are shown in Figs 5.23 and 5.24, first plain and then with numbered points, sixteen round the perimeter of each section. Piero's next three illustrations show

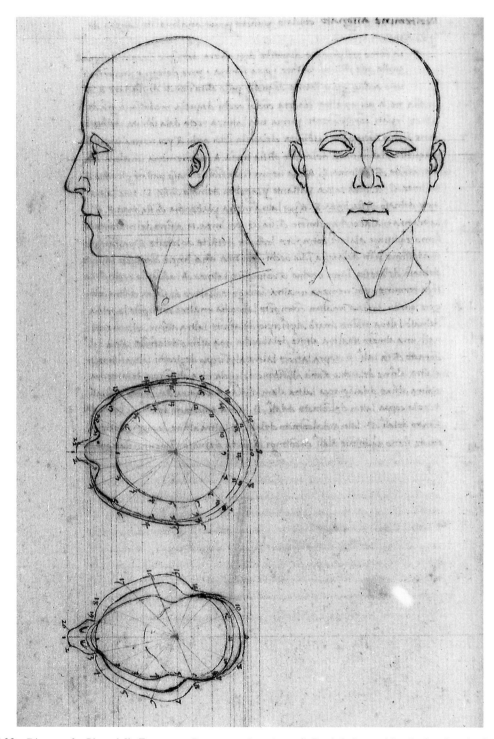

**Fig. 5.23** Diagram for Piero della Francesca, *De prospectiva pingendi*, Book 3, Proposition 8, showing the first set of drawings in the series for constructing images of a human head. From the manuscript in Parma (Biblioteca Palatina, no. 1576).

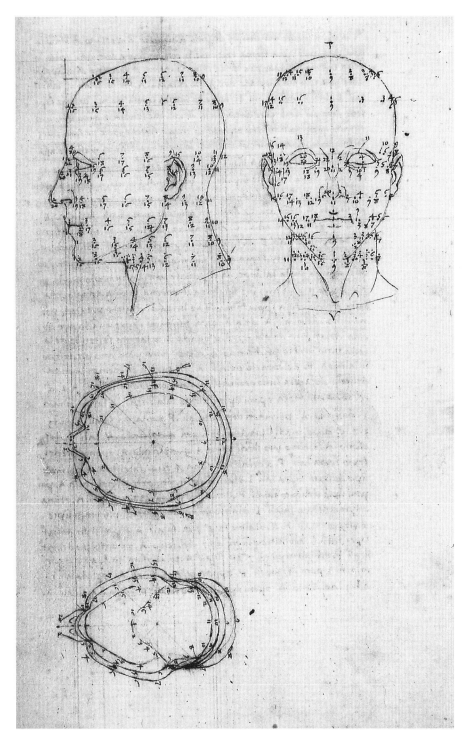

**Fig. 5.24** Diagram for Piero della Francesca, *De prospectiva pingendi*, Book 3, Proposition 8, showing the second set of drawings in the series for constructing images of a human head. The points shown in the first set of drawings have been given numbers. From the manuscript in Parma (Biblioteca Palatina, no. 1576).

the sight lines, that is the successive positions of the thread attached to the nail or needle at *A*. The viewpoint is slightly to one side and slightly below the head. The completed perspective drawing is shown in Fig. 5.25. It is accompanied by a retinue of rulers like those for the column capital but with their points more closely packed together.

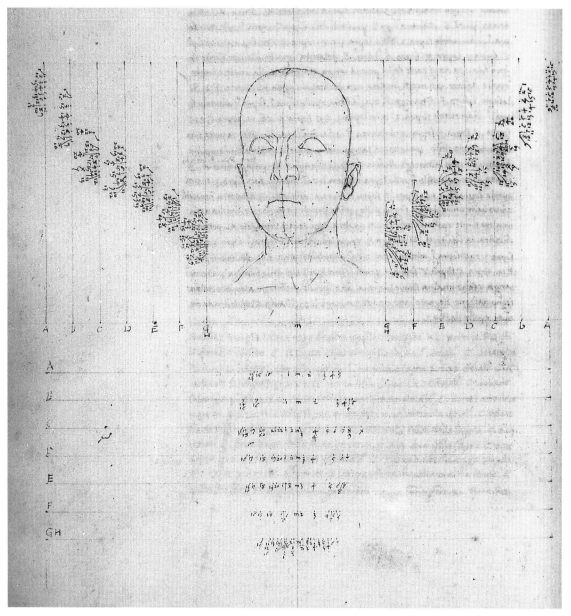

**Fig. 5.25** Diagram for Piero della Francesca, *De prospectiva pingendi*, Book 3, Proposition 8, showing the final drawing in the series for constructing the image of a human head seen slightly from below. The diagram includes the 'rulers' used in the construction. From the manuscript in Parma (Biblioteca Palatina, no. 1576).

This head bears a considerable resemblance to the head of the young man holding the Cross in the scene of *The Proving of the Cross* from the fresco cycle at Arezzo (see Fig. 5.26). In this scene, all the heads, and all the hands, and some details in the figure of the man who is being brought back to life by the Cross, show the characteristic lines of dots produced by the transfer of drawings to the plaster by means of the *spolvero* technique. It is accordingly clear that some preparatory drawings were made. The implication from Piero's perspective treatise is that the drawings were prepared by a method like that shown in our Figs 5.23–5.25. Computer analysis of the drawings on the wall might conceivably reveal whether this is possible. In any case, when one looks at the extreme delicacy with which many of the smallest details in the fresco have been painted—details that would be invisible to a viewer standing on the floor of the church—the labour that would have been involved in making perspective drawings does not seem disproportionate to the labour actually expended in the process of painting.

After the head shown in Fig. 5.26, Piero next takes lines of sight that will show the head much more steeply from below and more strongly tilted than before. The final version of the drawing is shown in Fig. 5.27, with its accompaniment of rulers. Again, a comparison can be made with a life-size head in a fresco, namely that of the second soldier from the left in the *Resurrection* that Piero painted in what was then the Town Hall of his native city of Borgo San Sepolcro (now called Sansepolcro); see Fig. 5.28.

The *Resurrection* is a solemn and imposing picture as well as a very subtle one. Analysis of its perspective is accordingly much easier when one is looking at a photograph rather than the original. Luckily, the elements involved are large enough not to be lost by the reduction in scale. It is immediately obvious that there is not enough information for any formal reconstruction of a perspective scheme. There appear to be no orthogonals to speak of. This lack of formality may be deliberate: the picture is some way above eye level and its position in what was the Council Chamber meant that it could and would be looked at from a wide variety of distances and angles. Relying on the sense of solidity imparted by the modelling in light and shadow of the figures, and the informal spaciousness of the landscape, may have been intended to ensure that the picture worked well in its particular visual context. In any case, Piero did allow for the fact that we are looking up at the picture: the sprawling soldiers seem to be seen from below and we are seeing only the underside of the rim of the tomb. However, Christ's foot, resting on the edge of the tomb, is seen from above, since its upper surface is visible. This implies an eye level something like that of the soldier on the right whose head cuts across the drapery falling from Christ's arm. None the less, there is no indication that we are seeing Christ himself from below. We seem, in fact, to be seeing Him straight on. Though the figure is convincingly three-dimensional, the non-naturalistic perspective gives it a weightless unearthly quality—which is perfectly in accord with the religious meaning of the scene. In technical terms, one can only say that the picture has been designed as having multiple eye levels. Its similarity to Masaccio's *Trinity* fresco is clearly not confined to the sense of monumentality, and an apparent fondness for the colour pink, but also extends to a willingness to bend the mathematical rules. For Piero—as for Masaccio—displays of virtuosity in the use of perspective should not be mistaken for a commitment to the belief that mathematically correct perspective is of overwhelming importance in the making of pictures.

All the same, it is clear that for Piero, as for Masaccio, the perspective is intended to look as if it were correct. Though the generalisation is dangerous, I think that on the whole this is true for Donatello also, anarchic though he seems to be in his use of mathematics. There are, however,

**Fig. 5.26** Piero della Francesca, *The Finding of the Crosses* (left) and *The Proving of the True Cross* (right), fresco, San Francesco, Arezzo. The figures are about life size.

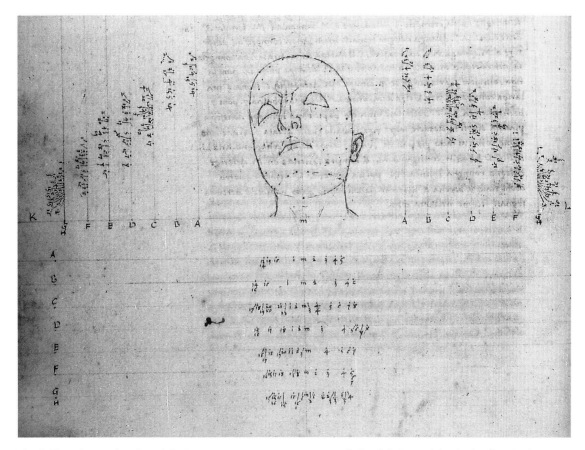

**Fig. 5.27** Diagram for Piero della Francesca, *De prospectiva pingendi*, Book 3, Proposition 8, showing the final drawing in the series for constructing the image of a human head seen more steeply from below and more strongly tilted than before. The diagram includes the 'rulers' used in the construction. From the manuscript in Parma (Biblioteca Palatina, no. 1576).

good artists who do not seem to care about apparent naturalism in quite the same way. One might look at Uccello as an example, but the clearest one seems to me to be Andrea Mantegna. A very striking instance of multiple eye levels is found in his *Virgin and Child with Saints Mary Magdalene and John the Baptist*, shown in Fig. 5.29. This picture probably dates from the last quarter of the fifteenth century. We have no reason to doubt Mantegna's proficiency in perspective—indeed he constructed quite a large number of pictures that are almost show-pieces for perspective. There can, however, be no doubt that in the altarpiece shown in Fig. 5.29 the figures of the flanking saints are seen from below while the central group is seen straight on. There appears to be no way of actually reconstructing the perspective, or rather perspectives, of this picture, but it is clear that we have two different eye levels.

To play tricks with perspective, as Mantegna does here, implies a degree of competence in handling the required construction. This kind of competence, or a convincing appearance of it, gradually became part of the painter's stock-in-trade. When printing of books with illustrations became reasonably commonplace, printed perspective treatises began to appear. They lagged a little behind learned geometry. Euclid was first printed in the 1490s, in Venice, but the first

**Fig. 5.28** Piero della Francesca, *The Resurrection of Christ*, fresco, Museo Civico (formerly Town Hall), Sansepolcro. The figures are about life size.

perspective treatise, that of Viator (a Latinised version of the name Jean Pélerin) was not published until 1505 (see Chapter 6 and Fig. 6.6 below). Most of these treatises are very simple, but they all follow something of the pattern laid down by Piero. Piero's treatise was never printed in the Renaissance—perhaps because it was too difficult—but it certainly circulated in manuscript and,

**Fig. 5.29** Andrea Mantegna, *Virgin and Child with Saints Mary Magdalene and John*, tempera on panel, 139 × 116.5 cm, painted between about 1475 and 1500, National Gallery, London.

like his work on algebra, was partly printed in the works of others. Piero's immortality as a mathematician has a large dash of anonymity on it. All the same, in the sixteenth century he was remembered as a mathematician rather than admired as a painter. Styles of painting had changed, making Piero's pictures look out of date, but correct perspective still mattered and he was recognised as having been a master of it.

# Chapter 6

## PRACTITIONERS AND PATRICIANS

Choosing to look exclusively at first class works of art has its disadvantages for the historian. Among other things, one tends to avoid seeing the snags inherent in a particular style. For the fifteenth and sixteenth centuries, excellent correctives are available, both in the form of second class works of art and in the form of unkind comments by artists about other people's work. Renaissance man, as revealed in the vivid pages of Niccolò Machiavelli's *The Prince* (written in 1513), is chiefly characterised by his determination to see off the opposition.

### Perspective and the ungrateful apprentice

One of the most competent practitioners of the style of painting fashionable in Florence in the late fifteenth century was Domenico Ghirlandaio (1449–1494). His skill was highly regarded in his time, so it no doubt seemed completely appropriate that he should be given the important commission to paint a cycle of frescos in the chancel of the church of Santa Maria Novella (Florence). One scene from the cycle is shown in Fig. 6.1. Ghirlandaio was working on the frescos from 1486 until 1490. Many of the participants in the scenes he painted are portraits of members of the family that commissioned the frescos, the Tornabuoni. For these figures, or at least for their heads, preliminary drawings would have been necessary. In fact some sheets of Ghirlandaio's preliminary drawings have survived, including two that not only show groups of figures but also give indications of the construction lines used in the perspective scheme. One of these drawings is shown in Fig. 6.2. As it is small in scale, this drawing is clearly a sketch of some kind. Full scale drawings would probably have been made to transfer some parts of the design to the wall (for instance the heads)—and the drawings, or copies of drawings, that were actually used for this purpose would, of course, have got both damp and dirty in the process, so it is hardly surprising that they have not been preserved. The sketches in effect merely confirm what we should probably have guessed by looking at the finished frescos, namely that Ghirlandaio planned his work carefully. The surviving more highly worked drawings are very close indeed to the corresponding parts of the fresco, suggesting that the planning was not only very careful but also very detailed.

This careful craftsmanship seems also to extend to the perspective schemes. The chancel is high, so the frescos are in three tiers, as were Piero della Francesca's frescos in San Francesco, Arezzo. Unlike Piero, however, Ghirlandaio has divided each tier into a series of separately framed scenes. And each scene has its own perspective. Each is convincing, so that as reproduced in books about the history of art the cycle looks highly satisfactory. Seen from the floor of the church, the effect is not comfortable. The individual spaces seem (to my eye) to open out like components of a chest of drawers. Nor is this reaction entirely anachronistic: Leonardo da Vinci complained about the

**Fig. 6.1**  Domenico Ghirlandaio, *The Birth of the virgin*, fresco, 1480s, chancel, Santa Maria Novella, Florence.

**Fig. 6.2**  Domenico Ghirlandaio, drawing, study for the fresco *The Visitation*, Galleria degli Uffizi, Florence.

multitude of perspectives found in some frescos. Naturally enough, he did not blame perspective but this particular use of it.

At the time he was painting in the chancel of Santa Maria Novella, Ghirlandaio's apprentices included the young Michelangelo Buonarroti. Presumably Michelangelo's father had chosen Ghirlandaio because of his high reputation. Michelangelo himself did not, it seems, have a high opinion of Ghirlandaio—at least, that is what he told people, including Giorgio Vasari, in later life. He even went so far as to claim that he had learned nothing from Ghirlandaio. The only painter from whom Michelangelo professed to have learned anything was Masaccio. This is credible in the sense that one can easily believe that the strong sculptural quality of Masaccio's work appealed to Michelangelo. Indeed, his opinion may well be what lies behind the fact that in praising the *Trinity* fresco Vasari speaks first of the figures. None the less, due allowance has to be made for the fact that by the mid-sixteenth century, when Vasari wrote his *Lives of the painters*, it was fairly usual for artists to wish to be seen as self-taught, strong-minded geniuses vastly superior to their predecessors in any respect you cared to name. Michelangelo is an arch-exponent of this mode. A pinch of salt may thus be required to accompany his story, retailed through Vasari, that one day Ghirlandaio was utterly amazed to find his talented apprentice was making a sketch of the scaffolding set up in the chancel. Scaffolding, since it consisted of straight pieces arranged at right angles, was a prime subject for a proper perspective construction. Ghirlandaio, in the Michelangelo version, is an old hack confronted with a young genius. The young genius is portrayed as embodying the mature

Michelangelo's dictum that the true artist 'should have his compasses in his eye', that is he should not need to resort to mechanical methods in order to get proportions correct. This is (of course) not to say that proportions do not matter, and some surviving working drawings by Michelangelo do include numbers to indicate how things are to be scaled up. His drawings for architecture sometimes include a full set of dimensions.

Nevertheless, one may be permitted a raised eyebrow. When the scaffolding was taken down after Michelangelo had painted about half of the ceiling of the Sistine Chapel (he worked there from 1508 to 1512), he realised that, although the painted architectural elements were satisfactory, the figures were too small. On the remainder of the ceiling the figures become a great deal larger, making the compositions much easier to read and giving them much greater dramatic impact (see Fig. 6.3). No doubt no-one dared to ask Michelangelo if he now wished to modify his remark about having compasses in the eye, or to tell him that Giotto would never have made such a mistake.

In any case, Michelangelo recognised the lesson. There is no sharp break, but the figures in the scenes in the second part of the ceiling are steadily increased in size until they reach satisfactory proportions. And in the *Last Judgement*, painted on the altar wall of the Sistine chapel about twenty years after Michelangelo completed the ceiling, the figures in the upper part of the picture have been made distinctly larger than those lower down, to allow for the fact that they will be seen from a greater distance (see Fig. 6.4). This effect is sometimes mentioned by writers on perspective. For instance Albrecht Dürer discusses it, in 1525, in connection with making the letters different sizes if inscriptions are to be placed at differing heights on a building (see Fig. 6.5).

Apart from the little local difficulty with natural optics on the ceiling of the Sistine Chapel, Michelangelo's art is not of a character that involves him in problems of perspective. In principle, his work as an architect might have done so (as we shall see, many architects did take a serious interest in perspective) but this was not unavoidable. Even detailed measured drawings may not always have been required. Antonio di Tuccio Manetti (1423–1497), Brunelleschi's first biographer, says that when Brunelleschi wanted to explain to his masons what shape some stones should be given, he made models of wax or soft clay, or carved the shape for them out of turnips (a large kind called 'goblets' that came into the market in winter). That was about a hundred years before Michelangelo; but even a hundred and fifty years after him, so the evidence suggests, Christopher Wren (1632–1723) made not drawings but wooden models of details to guide the masons working on St Paul's Cathedral in London. Even as an architect, Michelangelo may have been able to get away with being as uninterested in mathematics as he claimed to be.

## Perspective treatises for painters

In Michelangelo's time, however, most painters were interested in using mathematical means to obtain a convincing effect of the third dimension in their pictures. Competence in perspective seems to have become a normal qualification for painters, and the treatment of spatial relations is in general fairly naturalistic. Painters' needs for elementary instruction in perspective were met by a steady supply of printed treatises.

As mathematics, almost all of these treatises are dismal. Their examples are all essentially to be found in the treatise by Piero della Francesca, but there are many omissions and simplifications. In

**Fig. 6.3** Michelangelo, central part of the ceiling, fresco, Sistine Chapel, Vatican.

**Fig. 6.4** Michelangelo, *The Last Judgement*, fresco, Sistine Chapel, Vatican, altar wall.

some treatises there is a reference to the 'cone of vision'. In general, however, we start almost at once with the problem of drawing the image of a square. No proof or explanation is given. Following examples deal with a *pavimento* (sometimes more than one design appears), then with very simple solids, such as the cube, then, in a few treatises, we have simple architectural settings. Diagrams are supplied in the form of woodcuts, serviceable rather than elegant (an example is shown in Fig. 6.6), so the works have little visual charm to compensate for their lack of mathematical interest to the modern reader.

Most treatises stop after the simple solids. The more ambitious ones go on to what Piero called 'more difficult' bodies, but they do not use the method described in the third book of Piero's perspective treatise. Instead they propose the use of instruments, usually instruments like those used by contemporary surveyors, which work by constructing the image point by point.

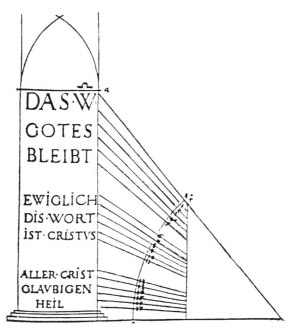

**Fig. 6.5** Albrecht Dürer, diagram showing change in size for lettering on a building. From his *Underweysung der Messung mit dem Zirkel und Richtscheyt* (Nuremberg, 1525), Book 3, Figure 28, sig. Ki *verso*.

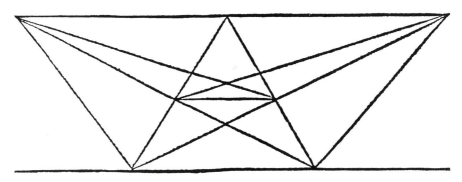

**Fig. 6.6** Viator, diagram to show distance points. From his *De artificiali perspectiva* (Toul, 1505).

One such instrument is shown in Fig. 6.7, which is taken from Albrecht Dürer's *Treatise on measurement with compasses and straightedge* (*Underweysung der Messung mit dem Zirkel und Richtscheyt*, Nuremberg, 1525). Here the problem is to construct the image of a lute—and, as can be seen, the actual presence of a lute is required. The sight lines, which we should probably more properly describe as eye beams, have been given the material form of string. The small ring fixed to the wall on the right, through which the string passes, represents the position of the eye. The further end of the eye beam is represented by the tip of the pointer held by the man on the left. The man on the right marks the position where the eye beam intersects the picture plane by adjusting strings stretched across the rectangular wooden frame fixed vertically to the table. This part of the instrument can be seen more clearly in the later (unmanned) diagram shown in Fig. 6.8. Once the cross-

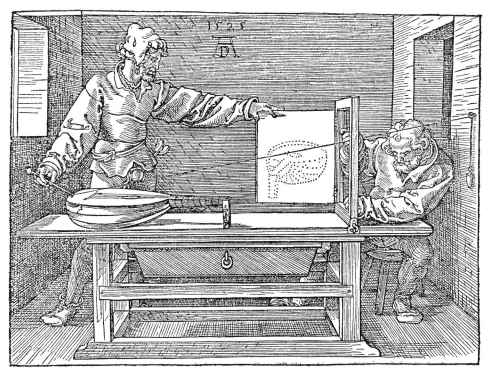

**Fig. 6.7**  Albrecht Dürer, illustration showing an instrument being used to draw a perspective image of a lute. From his *Underweysung der Messung mit dem Zirkel und Richtscheyt* (Nuremberg, 1525), Book 3, Figure 64, sig. Niii *recto*.

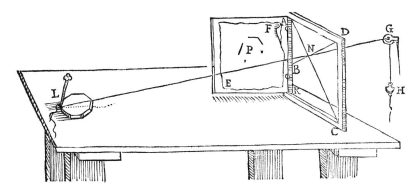

**Fig. 6.8**  Giacomo Barozzi (Vignola), ed. Egnazio Danti, *Le due regole della prospettiva pratica* (Rome, 1583). Danti's version of the instrument Dürer showed being used for drawing a lute (compare Fig. 6.7).

ing strings are in the appropriate position, the eye-beam string (*LNGH* in Fig. 6.8) is moved away and the panel with the drawing paper attached to it is swung into the picture plane so that the new point can be marked on it.

Another mechanical aid discussed by Dürer seems to go back to the early fifteenth century, namely the 'veil' (*velo*), which is described by Alberti and Leonardo da Vinci. In Dürer's version,

**Fig. 6.9** Albrecht Dürer, illustration showing a 'veil' being used to draw a perspective image of a naked woman. From his *Underweysung der Messung mit dem Zirkel und Richtscheyt* (Nuremberg, 1525), Book 3, Figure 67.

shown in Fig. 6.9, the veil has become coarse netting, which is used to provide something rather like a system of coordinate lines. Such squaring systems were familiar to artists as a method of scaling drawings up for transfer to the painting surface. An apparently later variant of the netting is shown in Fig. 6.10, where the frame with a movable sight is being used to obtain coordinates, and the corresponding points are marked on squared paper by a colleague. This engraving is taken from an Italian treatise on perspective, written by the architect Giacomo Barozzi (1507–1573), who is usually known as Vignola (from his place of birth). The treatise was edited and provided with a mathematical commentary by the professional mathematician Egnazio Danti (1536–1586). Unlike Dürer, Vignola shows his artists working from a statue. This may reflect the easier availability of suitable classical statuary in Rome as compared with Nuremberg. In any case, it illustrates the fact that artists were largely expected to learn by copying earlier art.

The instruments shown in Dürer's treatise include one which uses a very drastic method to obtain an accurate image. The method is to make a picture on glass, as shown in Fig. 6.11. The snag that the emerging picture progressively obscures the original is, one might have thought, discouragement enough for any artist proposing to use such a method—and the inconvenience of this would become even more apparent if the sitter showed a tendency to fidget. However, in the sixteenth century there would also have been the snag of the huge expense involved. Dürer says one requires a good piece of transparent glass. In fact, bubbles and striations were a continuing problem in even the highest quality glass of the seventeenth century. We know because astronomers regularly complain of the difficulty of obtaining suitable glass for telescope lenses. (In 1610, Galileo recommended Florentine rather than Venetian glass for the purpose.) Inadequate prisms led to difficulties in repeating Isaac Newton's optical experiments of the 1670s.

One rule of thumb when considering the sixteenth century is to remember that for certain purposes the rich were prepared to throw money at problems, so that in effect almost anything could be had for an appropriate price—the money paying for the time of skilled craftsmen and the wastage of materials in failed attempts to achieve the required standard. Thin glass was, of course, easier to make than the thick pieces required for lenses. All the same, Dürer's large sheet of glass seems unrealistic. It would have been much cheaper merely to employ an artist who could manage

**Fig. 6.10**   Giacomo Barozzi (Vignola), ed. Egnazio Danti, *Le due regole della prospettiva pratica* (Rome, 1583). Vignola's illustration of an instrument being used to draw a perspective image of a statue.

**Fig. 6.11** Albrecht Dürer, illustration showing a painting being done on glass. From his *Underweysung der Messung mit dem Zirkel und Richtscheyt* (Nuremberg, 1525), Book 3, Figure 63, sig. Nii *verso*.

without such help. Though he was clearly interested in obtaining a portrait showing what we should now call almost photographic accuracy and detail, Pope Julius II, the patron for whom Michelangelo painted the ceiling of the Sistine Chapel, got what he wanted by the simple expedient of issuing appropriate instructions to a good draughtsman, namely Raphael (Raffaello Santi, 1483–1520). The resultant portrait is shown in Fig. 6.12. Though Raphael presumably made preliminary studies, in the course of painting he changed his mind about the position of the head. The change can be seen in X-ray photographs. It indicates that this version of the portrait is the original rather than, as had been thought, a copy.

Changes of mind in the course of painting, usually known by the Italian term *pentimenti*, are a lot easier in oil paint, which is what Raphael was using in his portrait of the Pope, than they were in fresco or in earlier media used in panel pictures, in which the vehicle for the pigment was usually egg. This may account for another characteristically absolute dictum of Michelangelo's, to the effect that oil paint was for children and old women. Not much love was lost between Michelangelo and Raphael. As a sculptor, Michelangelo presumably liked the idea that a decision once taken was irrevocable: one cannot undo the effect of the chisel on a piece of marble. Painters in oils might make their work look equally decisive, but they did not actually have to be so in its

**Fig. 6.12** Raphael, *Portrait of Pope Julius II*, oil on panel, 108 × 80.7 cm, National Gallery, London.

execution. Whether as cause or effect, or mere coincidence, this went together with a more relaxed attitude to visibly mathematical elements in painting.

Indeed, in his treatise on perspective, called *The practice of perspective* (*La pratica della perspettiva*, Venice, 1568, 1569), the Venetian humanist Daniele Barbaro (1513–1570), who is addressing himself to patrons rather than to practitioners, complains that there is almost no perspective in painting now. This may seem highly unreasonable if we look at the realistic treatment of spatial relationships in Venetian art of Barbaro's time, for instance by Titian (Tiziano Vecellio, *c.* 1485–1576), see Fig. 6.13. However, reference to other texts shows that, by the time Barbaro was writing, to say a picture 'contained perspective' was to say it contained an architectural vista, that is, something like the deep colonnade on the title page of Vignola's treatise on perspective, shown in Fig. 6.14.

In these circumstances, it is clearly true to say that Titian's portrait of the Vendramin family (Fig. 6.13) has no perspective to speak of, though the receding orthogonals of the steps of the altar do help to provide a spatial setting for the figures. On the whole, Titian's use of architecture is not

**Fig. 6.13**   Titian, *The Vendramin family*, oil on canvas, 206 × 301 cm, probably *c.* 1543–1547, National Gallery, London.

extensive, and architectural elements are usually more important in establishing the pattern on the picture surface, that is the composition in the picture plane, rather than in constructing a pictorial space, that is in contributing to composition in space. A very good example of this is provided by Titian's *Madonna di Ca' Pesaro* (*c.* 1526), shown in Fig. 6.15. This picture is an altarpiece, and, in the Venetian manner, has life-size figures. The massive columns are used to emphasise the importance of the Madonna and Child, who have—for the first time—been placed off centre. This use of background architecture to emphasise foreground figures in the composition is not new, and it is not incompatible with using architecture to construct pictorial space. For instance, in Piero della Francesca's *The Proving of the Cross* (Fig. 5.26), the building seen as a background to the group of kneeling figures is used exactly in Titian's manner, without detriment to our reading of the spatial relationships concerned. To Daniele Barbaro and his contemporaries, the 'perspective' in Piero's picture would have been in the townscape visible on the right. As Barbaro remarks, this kind of use of perspective was not common in the paintings of the mid-sixteenth century. Titian's style is typical—and, indeed, much imitated.

Where perspective certainly was to be found in Barbaro's time was in stage sets. These often contained a suitable quantity of architecture, since Vitruvius' descriptions of ancient examples

**Fig. 6.14** Giacomo Barozzi (Vignola), ed. Egnazio Danti, *Le due regole della prospecttiva pratica* (Rome, 1583), title-page.

specified that townscapes should be shown in sets for both tragedy and comedy. The set for pastoral plays was to consist largely of landscape.

Though he is clear about what the sets showed, Vitruvius is rather vague about their style. However, twentieth-century scholars tend to agree with their Renaissance predecessors that ancient stage sets did involve the use of perspective. Unfortunately, the archaeological record has not, so far, helped to elucidate Vitruvius' text in this regard. It chiefly shows that the design of many theatres was significantly different from what he describes. For the Renaissance, however, Vitruvius was the sole and authoritative source on the matter, and it is very probably partly through his interest in the work of Vitruvius that Daniele Barbaro became interested in perspective. The other source of

**Fig. 6.15** Titian, *Madonna di Ca' Pesaro*, oil on canvas, 485 × 270 cm, *c.* 1519–1526, Santa Maria Gloriosa dei Frari, Venice.

Barbaro's interest has to do with the fact that like many of his well born contemporaries he had received an education that included a fair measure of mathematics. We shall deal briefly with this second source before returning to Barbaro's more specifically Vitruvian concerns.

## Mathematics in the education of the upper classes

As has already been indicated, mathematical skills were increasingly in demand in commerce and in the crafts from the fourteenth century onwards. In the early fifteenth century, it is not clear whether a higher level of mathematical education among artists should be seen as cause or effect of the use of new more complicated mathematical rules in the practice of their craft. In any case, most later artists seem to have applied the rules of perspective merely as rules, taking no particular interest in the underlying mathematics. This even seems to be true of Piero della Francesca's direct pupils, and of Raphael, whose father, Giovanni Santi (b. 1435–1440, d. 1494), had been court painter at Urbino and certainly knew something of Piero's painting. The printed treatises of the sixteenth century show that this attitude persisted. In this later period, fashions in painting did not change in a way that pushed artists towards learning more mathematics than before. Indeed, as we have seen, the contents of Piero della Francesca's perspective treatise seem to have proved more than adequate to the needs of the following century.

Other kinds of practical mathematics fared rather differently. The reasons are connected with political and, above all, military history. First there was military technology. Gunpowder had been known in Western Europe since the thirteenth century—its invention has sometimes been ascribed to the Franciscan friar Roger Bacon (1214–1294)—but it was only by the end of the fifteenth century that it became clear that cannon could consistently prove more dangerous to the enemy than to the user. Before then they tended to constitute an impressive threat rather than a useful weapon. Once cannon were established as a standard weapon, if still a monstrously expensive one, it became important to redesign fortifications, both to permit the use of cannon by a defender and to resist their use by an attacker. Numerous books on fortification appear in the early part of the sixteenth century. They are usually divided into two parts. The first describes how to design and build impregnable fortifications, the second part describes how to capture them. The one thing everyone was unambiguously agreed upon is that fortresses now needed to be a more complicated shape than before. The great invention of the military engineers is the polygonal fort, with polygonal bastions at its vertices. In surviving examples the masonry is usually so massive that the effect is grim and threatening rather than pleasingly geometrical. Curiously enough, a second invention, which is much less visually striking, was due to Michelangelo. When Florence was under threat in the 1520s, he was put in charge of the fortifications—after the authorities had thwarted his efforts to leave for somewhere safer. A sculptor was probably a good choice, since he was undoubtedly very good at visualising things in three dimensions. His innovation was, however, entirely prosaic, though very effective: to use earthworks, which absorbed the force of the cannonballs.

Some discussions of fortification introduce very elaborate polygons, but the commonest choice is the pentagon. A construction for a regular pentagon can be found in *Elements*, Book 4, but it is hardly the sort of construction one would wish to use in laying out the ground plan of a fortress. Nor do contemporary surveying instruments look particularly useful, since their angular scales were not very accurate. The practical method used by the Venetians, preserved in a file in the State Archives, is shown in Fig. 6.16. We are given the sides of the right-angled triangle. The triangle

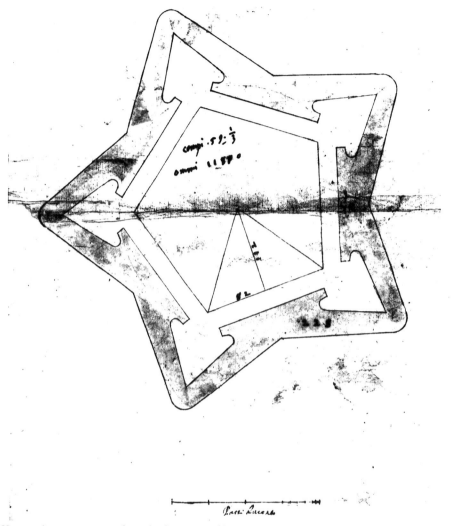

**Fig. 6.16**  How to draw a pentagon, from the Savorgnan file, Archivio di Stato, Venice.

can be constructed from rope, or by linking together the chains used by surveyors, and repetitions of it will then allow the complete pentagon to be laid out. The person who actually lays out the ground plan using this method does not need to be very good at mathematics, but it is clear that some mathematical expertise was needed to devise the method, and to calculate the lengths of the sides of the triangle.

It is even possible to suggest a name for the mathematician concerned. He may have been Niccolò Tartaglia (*c.* 1499–1557), who is now best remembered for his work on algebra and his pioneering book on the mathematics of gunnery *The new science* (*La nuova scienza*, Venice, 1537). Tartaglia was a personal friend of Giuliano Savorgnan, the man in charge of fortifications in the region of Venice, in whose file this method of drawing the pentagon was found. Savorgnan appears as one of the people setting problems in Tartaglia's *Various investigations and inventions* (*Quesiti e invenzioni diverse*, Venice, 1554), the book which contains Tartaglia's own, rhyming, account of the method of solving cubic equations that he had discovered. The rhyme was intended to be committed to memory.

The method concerned was first published by Girolamo Cardano (1501–1576) in his *Great art* (*Ars magna*) in 1546, with acknowledgement, in the preface, that the method was Tartaglia's. Cardano nevertheless got the credit for the method at the time, and there ensued what is politely called a 'priority dispute'. It involved a large amount of Renaissance-style character assassination as well as the exchange of algebraic problems that led to cubic equations. If one dodges the mudslinging, the controversy emerges as a rather interesting written version of the live problem-solving contests in which rival teachers engaged in order to advertise their skill to potential pupils. Unfortunately, in this particular case the historical truth failed to prevail, and Tartaglia's method of solving cubics still usually goes by the name of Cardano. Cardano's outliving his rival probably helped.

Tartaglia was luckier in regard to his work on gunnery. As seen from the twentieth century, the cubic equation still seems interesting, and the attempt at a mathematical theory of gunnery is hopelessly primitive, but, as we have seen, at the time Tartaglia's book was written gunnery was very important. Tartaglia did not in fact come up with anything that proved startlingly helpful, though his mathematical suggestions, for instance about the optimum elevation for the barrel, do seem to have been followed in practice. What is chiefly interesting about the book, for an understanding of the part played by mathematics in this period, is that it was Savorgnan, the military man, who asked Tartaglia, the mathematician, for help. Tartaglia tells us so in his preface. Savorgnan clearly had the idea that mathematics could be useful, and that some practical mathematical rules could be applied by his gunners. What we are seeing is the way that a certain amount of mathematical skill, and appreciation of the usefulness of such skill in others, was becoming part of the education of military commanders. Practical mathematics was infiltrating its way into the education of the upper classes.

## Vitruvian problems

Daniele Barbaro, whose treatise *The practice of perspective* (1568, 1569) has already been mentioned, was a fully paid up member of this patrician class. He seems to have received a good humanist education (the noted classical scholar Ermolao Barbaro (*c.* 1410–1471) was Daniele's great uncle) and this education had apparently included some practical as well as some Euclidean mathematics. As a well-educated patrician, he served as ambassador to the court of King Edward VI of England, and later as the leading Venetian representative at the Council of Trent. He held the high, though curiously titled, office of Patriarch Elect of Aquileia. In principle this office was ecclesiastical (and Barbaro seems to have been rather pushed into taking Holy Orders so that he could hold it) but it was also very important politically. Officials attached to the Papal Court in Rome, probably egged on by Milanese interests, engaged in a long-running attempt to undermine Barbaro's authority by casting doubt upon his religious orthodoxy. The truth of this is more difficult to unravel than that concerning the solution of cubic equations. In any case, there was no doubt of Barbaro's standing as a scholar. This aspect of him is emphasised in a portrait painted by Paolo Veronese (1528–1588), with whom Barbaro was personally acquainted. In this portrait Veronese shows Barbaro displaying two volumes of his edition of the work of Vitruvius (see Fig. 6.17), though the pages shown do not in fact correspond to actual openings of the books in question.

**Fig. 6.17**    Paolo Veronese, *Portrait of Daniele Barbaro*, oil on canvas, 121 × 105.5cm, Rijksmuseum, Amsterdam.

Barbaro's heavily annotated Italian translation of Vitruvius' *On architecture* was first published in 1556, in his native Venice. A second edition, with emended notes, appeared ten years later. In this same year Barbaro also published his edition of Vitruvius' original Latin text, again with heavy annotations. The 1566 editions became the standard texts of Vitruvius. Preliminary drawings for some of the architectural illustrations are believed to have been made by Andrea Palladio (1508–1580), whom Barbaro certainly knew personally, since they were both founder members of the Accademia Olimpica of Vicenza (to which we shall return) and Palladio was the designer of the Barbaro family villa (at Maser, on the mainland). The inside of the villa was painted with frescos by Paolo Veronese. The family clearly had a taste for employing first class talent.

In some ways, Vitruvius' *On architecture* can be seen as a practical text. It contains a great deal about the actual processes of building. In more learned style, it also contains theoretical disquisitions about the proportions appropriate to buildings using the various 'orders' of architecture (Doric, Ionic, Corinthian and Composite) and the buildings or parts of buildings to which each order is appropriate. Archaeological correctness in such matters was considered desirable in the sixteenth century. Following the Vitruvian rules ensured the building was beautiful—though there was plenty of room for argument when it came to designing churches, because Vitruvius (naturally enough) did not discuss them and a mere copy of a pagan temple would not have been either practical or acceptable. In addition to such matters of direct concern for building and building design,

Vitruvius also commented on various scientific subjects, himself exemplifying his own prescription that an architect needed to be a man of universal education. Among other things, this ideal architect needed to know some mathematics. Not all of this was elementary. We shall return in the next chapter to the problems associated with sundials—where the mathematics, be it said, seems to have been somewhat beyond Vitruvius' capabilities. Vitruvius is prudently brief on the subject of sundials. He is even briefer in his reference to the use of perspective in stage scenery. However, in both cases Barbaro responds with long and interesting mathematical notes. In fact, it looks as though what happened was that the notes in regard to perspective took on such proportions that something much shorter was put into the Vitruvius editions and Barbaro went on to write a separate treatise on perspective. So *The practice of perspective* is probably to be seen as a spin-off from Barbaro's work on Vitruvius. It also provided a mathematical supplement to Alberti's discussion of pictures in *On painting*. A new Italian translation of Alberti's work had appeared in Venice the year before Barbaro published his treatise on perspective.

Barbaro's version of Vitruvius' set for tragedy is shown in Fig. 6.18, which is in fact a copy of the picture given in the *Book on geometry and perspective* (*Libro di geometria e di prospettiva*, Paris, 1545)

**Fig. 6.18**  Daniele Barbaro, *La pratica della perspettiva* (Venice, 1569), after Serlio 1545, Vitruvius' stage set for tragedy.

by the architect Sebastiano Serlio (1475–1554). Architects' interest in stage sets was not only that they were expected to supply designs for the buildings concerned, drawing on their experience of providing patrons with pictures of proposed real buildings. In addition, architects were expected to be experts on stage design in its own right. After all, they were accustomed to having to give orders to carpenters and painters. Moreover, stage sets and stage machinery had much in common with the machines and temporary structures employed in building works. So Serlio's Vitruvian stage sets are not mere theoretical schemes. Serlio had practical experience of making such sets, which apparently consisted of a central panel and two pairs of side pieces, as shown in Figure 6.19. This diagram comes from the perspective treatise by Vignola published in 1583. It is one of the wood-cut illustrations provided by the editor and commentator Egnazio Danti. Barbaro has a very similar diagram, but it is less clear than Danti's. As can be seen, the orthogonals are put in by means of strings, one attached to a stout peg at the point *A*, which marks the position of the eye of the ideal observer. Mathematically, this method is a cop out, but it solved the practical problem.

No doubt some such method had been in use ever since the first introduction of perspective stage sets, which seems to have been in Florence in the 1480s. Descriptions and artists' sketches suggest that almost from the first there was a variety of designs. There were two main types. The first consisted of one large set, used as a single flat piece or, as in Barbaro's illustration, a combination of flats. An advantage of this type of design was that it allowed quick scene changes. One system is shown in Fig. 6.20, which is another of Danti's contributions to Vignola's treatise of 1583, and shows, in plan, his version of the movable system that Vitruvius described by the Greek term 'periaktoi' (περιακτοι). One third of a turn, simultaneously for each prism, changes the

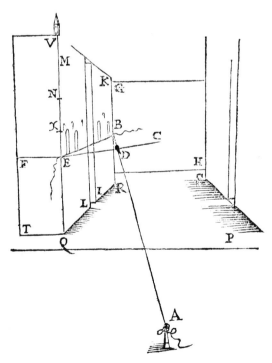

**Fig. 6.19** Giacomo Barozzi (Vignola), ed. Egnazio Danti, *Le due regole della prospettiva pratica* (Rome, 1583), p. 91, a stage set, showing how the perspective scheme was constructed.

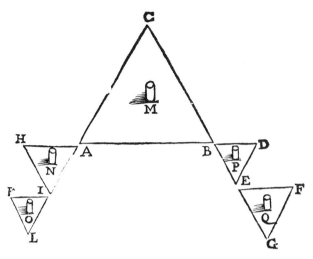

**Fig. 6.20**   Giacomo Barozzi (Vignola), ed. Egnazio Danti, *Le due regole della prospettiva pratica* (Rome, 1583), p. 91, quick-change stage scenery (Vitruvian 'periaktoi').

scene. In the late 1970s a system of this kind was introduced for changing the picture on advertising hoardings. The solid prisms of the early design ('prismavision') have now been replaced by hollow structures made of three juxtaposed slats ('ultravision').

The second main type of Renaissance stage set consisted of a screen of architecture running straight across the stage, with several openings affording more distant views, which could sometimes also be changed. An idea of the stage pictures this type of scenery produced is given by Veronese's painting *The family of Darius before Alexander*, shown in Fig. 6.21. This is the kind of

**Fig. 6.21**   Paolo Veronese, *The family of Darius before Alexander*, *c.* 1560(?), oil on canvas, 236.2 × 474.9, National Gallery, London.

ornate spectacle that we should now associate with the relatively slow-moving dramatic form of opera rather than with an ordinary play. Stage spectacles of the sixteenth century were, indeed, often partly accompanied by music or supplemented by musical interludes between the acts. Opera proper—defined as 'a new form of entertainment in which the main characters sing all the time'—was introduced by the author–musician team of Alessandro Striggio (1573–1630) and Claudio Monteverdi (1567–1643) in 1607, with the first performance of their *Orfeo* in Mantua on 24 February. Monteverdi's definition of what was new about this piece is the one just quoted. The first public opera house opened in Venice in 1637.

## The Teatro Olimpico

Plato had taught, it was said, in a grove of trees outside Athens called by a name which was latinised as *academia*. In Italian this became *accademia*, and in the sixteenth century the name was taken up by many associations which saw their purpose as that of cultivating learned activities practised by the ancients. One of the first of these was the Accademia Olimpica of Vicenza, founded in 1551. As we have already mentioned, its founding members included Daniele Barbaro and Andrea Palladio.

Palladio had begun his career as a stone carver. It was thanks to an early patron, Count Giovanni Giorgio Trissino (1478–1550), a leading intellectual of Vicenza, that he acquired a classical education, and the classical-sounding surname Palladio to go with it. Palladio's original name was Andrea di Pietro della Gondola.

One of the purposes of the Accademia Olimpica was to build a Vitruvian theatre in which to stage an ancient tragedy. This was accomplished in 1585, with the staging of Trissino's Italian version of Sophocles' *Œdipus the King*, the play Aristotle cites as the perfect exemplar of Greek tragedy. The theatre built for the purpose, and inaugurated by this production, was named after the institution that owned it and is accordingly called the Teatro Olimpico. The architecture, and the front part of the stage set, is by Palladio. To prove they had an ear as well as an eye for first class talent, the Accademia Olimpica commissioned the music for *Œdipus* from the First Organist at St Mark's, Venice, the famous composer Andrea Gabrieli (*c.* 1510–1586). The budget for this production appears to have been of a size comparable to that of the most spectacular of Hollywood spectaculars.

The exterior of the Teatro Olimpico is mainly blank wall, with the neat marble window and door frames that are standard in the region of Venice. (The greyish marble of which they are made came from Istria, a region on the East coast of the Adriatic, then controlled by Venice.) Two views of the interior of the auditorium are shown in Figs 6.22 and 6.23. Since scale matters so much in one's apprehension of architecture, the photographs will look very inadequate to anyone who has seen the actual building. They fail to convey either the monumentality of the space or the sense it gives of being on a human scale. This latter effect is partly due to Palladio's care over detail.

The design of the Teatro Olimpico was inspired by and largely taken from Vitruvius' description of ancient theatres. There are, however, notable divergences from Vitruvius. One can be seen in Fig. 6.23. In Vitruvius, and in all classical theatres found by archaeologists, there is a complete separation between the tiers of seats and the stage area. Palladio, one of whose leading characteristics as an architect was to create unified spaces, has folded the stage scenery round to join it to the seating. Though the change may seem trivial when described in this way, the result is a space which feels very different from that of a true classical theatre. It is not known whether this departure from Vitruvius was recognised as such by Palladio or by the other Academicians.

Another departure, visible in the plan of the theatre shown in Fig. 6.24, must have been entirely deliberate. Vitruvius describes the seating area as semicircular. Palladio has made it semi-elliptical. As luck would have it, the change of shape is easy to explain. It is known from surviving documents that at the time the theatre was designed the Accademia Olimpica did not own the part of the site that is now taken up with the perspective scenery behind the main architectural screen at the back of the acting area. So there was not room to make a semicircular theatre. The remainder of the site was bought after Palladio's death. Presumably it was then decided that his design for the main part of the theatre should be retained and the additional space used for the extended scenery. Although it is possible that this scenery is based, at least partly, on designs by Palladio, the story of the site makes it certain that the final form of the scenery is due to someone else. It is probably by Palladio's most talented pupil Vincenzo Scamozzi (1552–1616). Palladio's design for the architectural screen may have been modified in order to accommodate the new scenery, for instance by enlarging the central arch to allow a better view of the three streets seen through it.

The scenery shown in the photographs in Figs 6.22 and 6.23, and in the elevation and section in Figs 6.25 and 6.26, is the original scenery for the production of Œdipus in 1585. It was so much

**Fig. 6.22** Interior of the Teatro Olimpico, designed by Andrea Palladio, a view to show the *frons scenae*.

admired that it was left in position. The city it shows is Thebes, 'of the seven gates' (according to Homer), which is why we have seven streets, some with city gates visible in the distance. This scenery is not painted flats but is made up of three-dimensional pieces, each consisting of one or more buildings. There does not appear to be a single consistent perspective scheme. Instead, the buildings of each street are constructed to look right when seen from the particular part of the

**Fig. 6.23** Interior of the Teatro Olimpico, designed by Andrea Palladio, a view to show the join between the *frons scenae* and the seating.

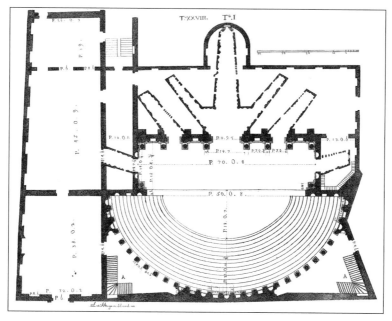

**Fig. 6.24**  Andrea Palladio, Teatro Olimpico, plan, from Bertotti-Scamozzi, 1790.

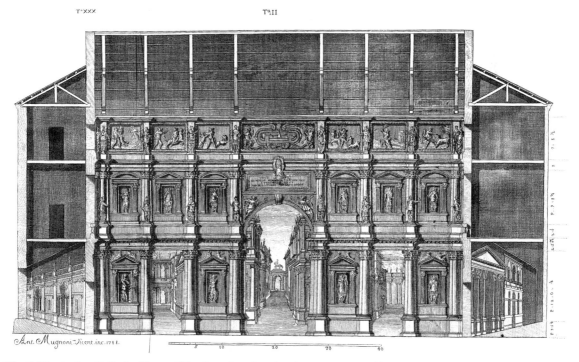

**Fig. 6.25**  Andrea Palladio, Teatro Olimpico, elevation showing stage screen (*frons scenae*) and scenery, from Bertotti-Scamozzi, 1790.

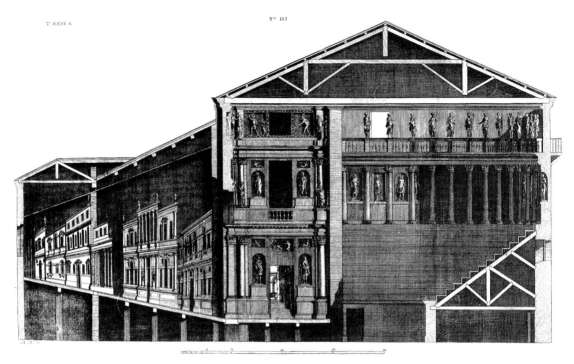

**Fig. 6.26**   Andrea Palladio, Teatro Olimpico, section, from Bertotti-Scamozzi, 1790.

auditorium from which they are visible. Only part of the scenery is visible from any one position. The most expensive seats are, naturally, those in the centre block, having the view through the central arch.

The engraved drawings of the theatre shown in our figures were published in the late eighteenth century by an admirer of Scamozzi's architecture, whose esteem even led him to add Scamozzi's surname to his own. It seems likely that they are too idealised to allow any investigation of the sight lines. Their scale is certainly too small to permit any calculation of a viewing distance.

Actually visiting the stage produced some surprises. No doubt it should not have been a surprise, except to one dazzled by the effect when seen from the auditorium, to find that the heavy carpentry of the hidden parts had the ancient look of shipwrights' work that one also finds in the roofs of churches. The technology, as it were, provided a reminder of the age of the art. A further reminder was provided by the small metal plaques attached to the backs of some pieces. At the top of each was a hook for a lamp. The metal was protection for the wood, and a record of the exact positions of the lighting used in 1585. Much of the lighting had come from within the set, and it explained some otherwise strange features. For instance, the fact that they were back-lit no doubt partly accounted for the very crude shaping given to the columns on some of the houses in the central street. Properly, the perspective should have given them an elliptical cross-section. The actual shape was more or less as shown in Fig. 6.27, that is rather like the head of a modern broom. The effect of curvature was almost entirely due to the illusionistic use of paint. This was also true of the detailed modelling of such things as the decorative mouldings on the house fronts. However, the

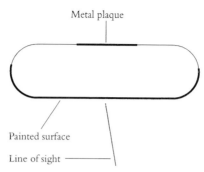

**Fig. 6.27** Sketch of part of the scenery of the Teatro Olimpico, showing the cross-section of a column in the vista of the main street, indicating the extent of the painted surface and showing the position of the metal plaque that protected the wood from the flame of the lamp.

statues were actual statues, free standing and finished all round. Presumably it had been easier to do them that way than to take detailed account of possible lines of sight.

Though less complete than the sculptures, the houses were also usually finished in parts to which the audience's eye could not reach—that is, they were neatly painted, mainly to look like marble, in parts from which I could not see into the auditorium. Since the scenery had been removed for safe keeping during the 1939–1945 war, one cannot be certain it is now exactly in its original position, but it is clear that the original designs had allowed a considerable margin of safety. Indeed, the margin was so large—a matter of more than a metre in places—that it suggested the pieces must have been built and painted elsewhere before being assembled on stage and lined up correctly. The alignment was no doubt checked with string, as shown in Fig. 6.19, and the overall shapes of the pieces had presumably been arrived at in the same way.

The actual shapes of the pieces were, in principle, frusta of oblique pyramids on rectangular bases, the ornamental mouldings being added on as frusta of oblique triangular pyramids. When confronted with such shapes my visual system resorted to sending back messages like 'input defective, refuse to process'. Knowing mathematical names for the shapes was no help at all, and it seemed most unlikely the carpenters could have been given their orders in the form of drawings. Some version of the Brunelleschian turnip technique would have been much more useful. Since I am not fond of turnips, my own version involved pieces of cheese.

Not surprisingly, having inspected the means used in the illusion in no way weakened the illusion itself. The conditions are, in fact, almost ideal for such an illusion, because the audience cannot get close enough to exercise any critical faculties it may imagine itself to possess. However, the rather easy access to the back stage area suggests that it may have been part of the original intention that members of the audience should be allowed to inspect the scenery from close quarters. Only a very few of the passages I used could have been intended for access by actors, since entering anywhere but at the very front will make an actor look over-size. The scenery is merely scenery, not a possible extension of the main acting area, which is in front of Palladio's architectural screen. So the numerous passageways through the scenery were probably for visitors. The upper classes were expected to be interested in how the illusion worked. It was, indeed, for such people that Daniele Barbaro had written his book about perspective. Their education had included a fair amount of mathematics, some of it of the ostensibly useful variety associated with *perspectiva-*

based crafts such as surveying. The questions such people asked sometimes elicited mathematically interesting answers, particularly when they were addressed to competent professional mathematicians—as they apparently often were, since princes and other grandees increasingly tended to employ such people as permanent members of their entourage. The effects of professional mathematicians turning their attention to their patrons' mathematical questions will be discussed in the next chapter.

# Chapter 7

## THE PROFESSIONALS MOVE IN

The very fact that someone of Daniele Barbaro's social standing should write a treatise on perspective makes it clear, as do the actual contents of the treatise, that mathematical matters such as the use of perspective in paintings and stage sets were considered suitable topics of conversation in patrician circles as well as in learned ones. There is much other evidence to confirm that this was so, though Barbaro's treatise was also read by practitioners: for instance, Diego Velázquez (1599–1660) owned a copy of it. Another part of mathematics suitable for conversational discussion was astronomy. This too has a connection with art, since astronomical subjects, often with mythological or astrological connections, are a regular component of schemes of decoration, usually, but not always, on ceilings. For instance *The origin of the Milky Way*, shown in Fig. 7.1, was painted by Jacopo Tintoretto (1518–1594) for the Holy Roman Emperor Rudolf II (reigned 1576–1612) in the late 1570s. The picture was apparently never delivered, possibly because Tintoretto was unwilling to part with it before payment had been received. Rudolf's treasury tended to be extremely slow in paying money he had promised. In any case, Rudolf's interest in astronomy went well beyond works of art. Holders of the office of Imperial Mathematician at his court included two of the most distinguished astronomers of the day, Tycho Brahe (1546–1601) and Johannes Kepler—both of whom had difficulties in arranging for payment of their salaries. Both found their duties included lengthy consultations on matters of astrology, frequently in connection with political questions, such as whether another Turkish invasion should be expected. This application of astrology was generally regarded as less reliable than its use in medicine or in weather prediction, but Rudolf believed in it.

### Calendar reform

Together with astrology, another normal area of upper class interest in astronomy was calendar reform. This had appropriate patrician and ancient connections: Julius Caesar (101–44 BC) had given his political support, and his name, to a reform of the calendar. Previous Roman, and indeed Greek, calendars had been highly complicated, partly at least because rulers of one sort and another frequently saw fit to add extra holidays here and there. Astronomers (who naturally preferred all years to be the same length) seem to have mainly used the Egyptian calendar, in which all years were 365 days long. For civil purposes the year had twelve months of 30 days each and there were five extra days to make up the number. The price for this simplicity was that the seasons did not stay fixed to particular dates. Meteorological differences between seasons are not as marked in Egypt as in many other countries, but one seasonal event was of overwhelming importance: the flooding of the Nile. As centuries passed, the time of the flood moved slowly from month to month. So too did the dates at which certain stars were visible at night. The Egyptians seem to

**Fig. 7.1**  Jacopo Tintoretto, *The origin of the Milky Way*, painted about 1576–1580, oil on canvas, 148 × 165 cm, National Gallery, London. A piece has probably been cut away from the base of the picture. The missing part is believed to have shown a recumbent female figure representing Earth.

have taken the view that it would all come back to where it started in 1461 years, so that was all right. The actual date of the Nile flood was predicted by astronomical observation. When the bright star we call Sirius just became visible in the Eastern sky before sunrise, there was about a fortnight to go before the onset of the flood, so seeds should be planted. This system worked. All other systems used in Europe and the Near East have attempted to connect the civil calendar to the apparent motion of the Sun, thereby fixing it to the seasons of the year, and have sometimes also allowed the Moon to define months, thereby asking for trouble by obtaining months that do not fit exactly into a year. Julius Caesar took advice from astronomers and imposed his compromise version on the Roman Republic in the year 707 A.U.C. (Ab Urbe Condita, from the foundation of the city), that is 46 BC.

By the fifteenth century, the Julian calendar was showing its age. Enough time had elapsed for the dates of astronomical events to have moved noticeably away from their positions in the calendar at the time of its introduction. The one that attracted attention to itself was the Spring equinox, that is, the moment in the year when the Sun crosses the celestial equator from South to North. (This event shows up in day and night then being of equal length everywhere on the Earth, hence the name 'equinox'.) The importance of the Spring equinox was its connection with the date of Easter. The Last Supper of Christ and his disciples was a ceremonial meal associated with the Jewish festival of Passover. Accordingly, the date of Easter was derived from the method of fixing the date of Passover in the Jewish calendar (which is luni-solar), and involved finding the date of the first full Moon after the first Sunday after the equinox. Full Moon is quite easy to see. The equinox is not. None the less, uncertainties about which full Moon gave one Easter were considered embarrassing. Astronomers were encouraged to worry about the date of the equinox, and to give their advice on redesigning the calendar so that the Spring equinox would once again be fixed where Julius Caesar had thought to fix it, on 21 March. Tinkering with the pagan contribution to the problem somehow seemed preferable to tinkering with the Jewish one—perhaps because the latter was taken to be Christian?

The ecclesiastical connection of the calendrical problem may make it seem appropriate that the eventual reform was brought in by the then Pope, Gregory XIII. He decreed that ten days would be omitted after 4 October 1582, and introduced a modified system of leap years designed to stop the Spring equinox getting more than a day away from 21 March. Good Catholics, and level-headed astronomers of other persuasions, at once accepted this system. In other quarters, it was slower to impose itself.

As has been hinted, there was a political element in this. Gregory XIII had won a race. The other runners had included the Grand Dukes of Tuscany, who, as members of the Medici family, claimed descent from Julius Caesar. Portrait busts indeed show them looking as Roman as possible. Their wish to emulate the glory of their putative ancestor certainly has some connection with the astronomical activities of one of their court mathematicians, Egnazio Danti. Danti had been a professor at the University of Bologna (in the Papal States), and after leaving the Medici service took employment as a mathematician at the papal court in Rome (where he was duly consulted about the proposed calendar reform). We have already mentioned his edition of Vignola's treatise on perspective (see above and Figs 6.10, 6.14, 6.19 and 6.20). In Florence, his activities seem to have been largely connected with learned mathematics, in the form of astronomy. He is known to have made, or more probably supervised the making of, a number of astronomical instruments, some apparently of his own design. He also made series of observations of the Sun. Two observing instruments that he used are to be found attached to the façade of the church of Santa Maria Novella (Florence), which happens to face almost due South.

Two larger observing instruments can be found inside the church. In the early 1570s Danti made a hole in the round window (*occhio,* literally 'eye') in the façade of Santa Maria Novella to enable him to use the church as a *camera obscura* for making observations of the height of the Sun at noon. This kind of set-up is usually called a *gnomon*. The principle involved is shown in Fig. 7.2. Later, Danti made another hole, higher up, in the strip of dark marble running up from the *occhio*. To admit light into the body of the church from this higher aperture, it was necessary to cut a slot in the vault of the church. This slot is still visible. The positions of Danti's astronomical instruments on the façade of the church and the locations of the apertures for the two gnomons are

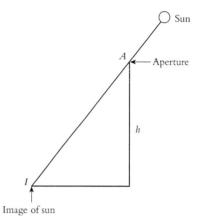

**Fig. 7.2** The principle of a gnomon, that is, a set-up that constitutes a *camera obscura* for observations of the Sun.

shown in Fig. 7.3. When restoration work was done on the *occhio* in the 1970s, one small piece of stained glass in a suitable position in the window was found to be much newer than its neighbours, and was consequently identified as the location for the aperture of the lower gnomon. It was decided that a replacement for the aperture should be put in place. Figure 7.4 shows the image of the Sun that it gives on the floor of the church.

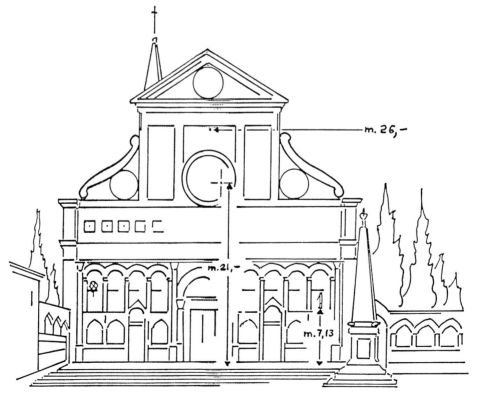

**Fig. 7.3** Façade of Santa Maria Novella (Florence), showing the positions of Danti's quadrant and equatorial armillary, and the apertures for the two gnomons. Reproduced by courtesy of Prof T. B. Settle, who rediscovered Danti's gnomons.

**Fig. 7.4** Image of the Sun from the reconstruction of Danti's lower gnomon (aperture in *occhio*), in Santa Maria Novella (Florence) a few days before the spring equinox in 1986. Dimensions are in cm.

## Editing perspective: Egnazio Danti

As a professional mathematician with an interest in observational astronomy, Danti was naturally also concerned with optics. He made an Italian translation of some Euclidean optical works (*La prospettiva di Euclide*, Florence, 1573) which seems to have been widely read, probably partly because Danti supplied extensive and learned commentaries on the texts concerned. By this date, works on optics (other than editions of ancient or medieval texts, of course) were generally expected to include something about perspective. Danti's interest in the matter was, however, much more deeply rooted: he came from a family of practising artists. Its most prominent member was his elder brother, Vincenzo Danti (1530–1576), a sculptor, architect, poet and writer on the theory of art who was a member of the Florentine Accademia del Disegno. Vincenzo Danti worked mainly in Florence, where his bronze group *The death of St John the Baptist* is still in position over one of the doors of the Baptistery (see Fig. 7.5). From his earliest years, Egnazio must have been familiar with the manuals on perspective, manuscript and printed, that were in use in artists' studios. It is therefore not surprising that in his later years he should have undertaken the task of editing for publication the treatise on perspective written by Vignola (for its title page see Fig. 6.14). Danti's introduction to the work includes a brief but very interesting history of perspective:

Which [perspective] it is clear that they [the ancients] knew about, if we examine what various authors write about this Art, but (although research has been diligent) we know of no book or written document which

**Fig. 7.5**   Vincenzo Danti, *The death of St John the Baptist*, bronze, baptistery of the cathedral, Florence.

has come down to us from ancient practitioners, although they were most excellent, as is convincingly shown by the descriptions of the stage scenery they made, which was much prized both in Athens among the Greeks and in Rome among the Latins.

The mention of ancient stage scenery is presumably a reference to Vitruvius, but it has connections also with the stage practice of Danti's own time, which, as we have seen, was consciously attempting to revive the classical style. After this humanist preamble, Danti turns to the moderns:

But in our own time, among those who have left something of note in this Art, the earliest, and one who wrote with best method and form, was Messer Pietro della Francesca dal Borgo Sansepolcro, from whom we have today three books in manuscript, most excellently illustrated; and whoever wants to know how excellent they are should look to Daniele Barbaro, who has transcribed a great part of them in his book on Perspective.

There follows a list, almost amounting to a complete international bibliography, of perspective treatises addressed to artists, on some of which Danti makes unkind but not completely inapposite comments. For instance, he correctly describes Viator's very elementary work (*De artificiali perspectiva*, first edition Toul, 1505, see Fig. 6.6) as having 'a greater abundance of pictures than words'.

His tendency to be dismissive of older works, such as Dürer's *Treatise on measurement...* (*Underweysung der Messung ...*, Nuremberg, 1525, see above and Figs 6.7, 6.9 and 6.11), is probably due to his making insufficient allowance for the higher general level of mathematical education prevailing in his own time compared with theirs. That is, Danti is not quite fair as a historian. His mathematical comments, however, make a refreshing change from many twentieth-century historians' over-enthusiastic search for evidence of mathematical 'progress' among perspective treatises. As Danti implies, such progress seems largely illusory to anyone who has taken even a moderately careful look at the perspective treatise written by Piero della Francesca. Danti's remark about Piero's treatise being available in manuscript is a reminder that in the sixteenth century printed books had not yet entirely ousted handwritten ones. Possibly one should not make too much of the fact that Piero's treatise never found its way into print in its own right.

As we have seen in the last chapter, Danti provided a number of additional illustrations to Vignola's treatise *The two rules of practical perspective*. In fact, he also made very substantial additions to its contents, by providing detailed mathematical commentaries where Vignola had merely adopted the usual practice of describing the rules to be followed. Different typeface is used to distinguish Danti's commentaries from Vignola's original text, and Danti's additional figures are woodcuts, making them easily distinguishable from the much more elegant copper engravings supplied by Vignola. All the same, it is not always clear how far the editorial work has gone. We can be sure, though, that the long and thorough discussion of optical preliminaries is by Danti. When we switch to Vignola, we find that—taking his narrative cue from Homer and Virgil rather than Euclid—he begins by plunging into the middle of things. The first diagram looks very like the diagram from Viator shown in Fig. 6.6. However, in Vignola's version this diagram turns out to be showing a purely mathematical result: that if a triangle is drawn between two parallel lines, the base along one line and the apex on the other, and from two points on the upper parallel equidistant from the apex there are drawn lines to the two ends of the base, then a line through the points of intersection of these lines with the sides of the triangle will be parallel to the base. There follow some further explorations of the mathematical properties of figures involving triangles. All the diagrams concerned have an air of having been extracted from treatises on perspective. The mathematics is quite easy, but occasional use of terms that seem to relate the theorems and diagrams to perspective only serves to increase one's feeling that one has somehow become engaged in a long parenthesis where one had expected to find a beginning. This style is probably an indication that Vignola (or Danti, if he has re-ordered the material) is addressing himself not to a complete novice but rather to someone who has already seen diagrams like these in perspective treatises. We have reached page 50 before we encounter the heading 'The First Rule' and learn at last what the treatise is about:

## That one can proceed by various rules. Chapter I.

Although many have said that in Perspective only one rule is true, condemning all the others as false, in what follows we shall show that one can proceed by various rules, and draw in correct Perspective. We shall consider the two principal rules, from which all the others depend, and while it happens that they look different in use, they none the less all arrive at the same result, as will be demonstrated clearly with good arguments. And first we shall consider the [rule that is] more well known and easier to comprehend, but longer and more tedious in application; and in the second [part] we shall consider the [rule that is] more difficult to comprehend, but easier to follow [in practice].

This is the complete text of the chapter. In mathematical terms, Vignola is right. But his reasoned approach, while widely read, does not seem to have settled the matter. There is much evidence for claims and counterclaims concerning the correctness or otherwise of the perspective in various works of art that some artists and patrons found, or professed to find, visually unsatisfactory. And the wide acceptance of the name 'legitimate construction', coined to describe the Albertian construction in 1625 (see Chapter 2), strongly suggests that many people continued to believe that only one rule could be correct. The harsh and uneasy intolerance in matters of religion that characterises this period seems to have its milder counterpart in this bit of mathematics. Perhaps casual readers were inclined to take Vignola's arguments as merely attempts at peacemaking rather than mathematics?

As we read on in Vignola's text, it becomes clear that the 'first rule' is that for the Albertian construction (shown in our Figs 2.4 and 2.5), and the 'second rule' is that for the 'distance point construction' (shown in our Figs 2.8, 2.9, 2.10 and 2.12). Vignola seems to be at pains to make the two construction methods look as similar as possible in his diagrams—presumably because he is maintaining that they do in practice give the same results. In fact, the most noticeable difference between the diagram for the first rule and that for the second is that the first shows the observer as a fully clothed woman standing at the left, whereas the second shows the observer as a naked man (with a bit of drapery over one arm in the best classical manner) standing at the right. Reading the remainder of each of the diagrams is complicated by the use of 'folding'. As a result of this technique, elements from the picture plane, and from the ground plane, are brought into the plane of the vertical section through the observer (that is the ostensible plane of the diagram). The finished diagram is thus a combination of elements from three planes that in reality are mutually perpendicular. These representational conventions would have been familiar to sixteenth-century readers, but the twentieth-century reader may find it useful to turn to the accompanying text for clarification. Changing drawing conventions have reversed the proverbial situation and in these particular circumstances two hundred words do much better than a picture.

In any case, Vignola is not short on words, or on multiplying examples in the usual style of a practical handbook. Danti's notes add mathematical weight. However, as we have seen in his discussion of stage scenery (see Chapter 6), Danti is also concerned with practical solutions to mathematical problems. He gives full descriptions of a large number of devices used as aids to drawing in perspective. Some of Danti's text is almost certainly addressed to interested patrons rather than actual practitioners, but the humanist style takes the form of references to ancient mathematical learning rather than to ancient art or ancient texts about art. It was probably partly as a result of its up-to-date mathematical professionalism, as well as Vignola's fame as an architect and practitioner of perspective, that Danti's edition of Vignola proved so successful. It went through many editions and seems to have been widely read.

## Editing the ancients: Federico Commandino

At the very end of his list of previous writers on perspective, Danti refers to some work on the subject by Federico Commandino (1509–1575). This, though Danti does not say so, is on a very different level from the other books he has mentioned.

Commandino, whose university training (at Padua and Ferrara) was in medicine, is best remembered now, and indeed was best known in his own lifetime, as a learned humanist whose knowledge of ancient languages was allied with a very considerable mathematical competence. This combination of talents made him the foremost editor and translator of Greek mathematics. His publications of the works of Archimedes (1558), Apollonius of Perga (1566) and Pappus (1566, 1588) at once became the standard editions. These publications, though they were indeed the product of humanist historical scholarship, were not produced chiefly because of the historical importance of the texts concerned. At the time they were printed, the texts were also of much intrinsic mathematical interest. By making them available to mathematicians, Commandino's editions in fact exercised considerable influence over the development of mathematics in the later sixteenth and seventeenth centuries. For instance, when, in the years between about 1604 and 1615, Johannes Kepler invented what was later called the 'calculus of indivisibles' (the immediate ancestor of the infinitesimal calculus), he justified his use of approximate methods involving very small magnitudes by referring to Archimedes' use of a similar kind of procedure in finding the area of the circle (in a treatise called *The measurement of a circle*). Detailed references make it clear that the edition Kepler had been using was that of Commandino.

Commandino engaged with perspective through his study of the *Planisphere* (*Planisphærium*) of Claudius Ptolemy. This work is about stereographic projection, a form of conical projection in which the images are those of points and circles on a sphere and the centre of projection itself lies on the sphere. This form of projection is that most commonly used on medieval astrolabes, for which the sphere concerned is the celestial sphere and the centre of projection is usually the South celestial pole, while the plane onto which the images of stars and various celestial circles are projected is that of the celestial equator (see Fig. 7.6). Under stereographic projection circles remain circles. Thus the circle

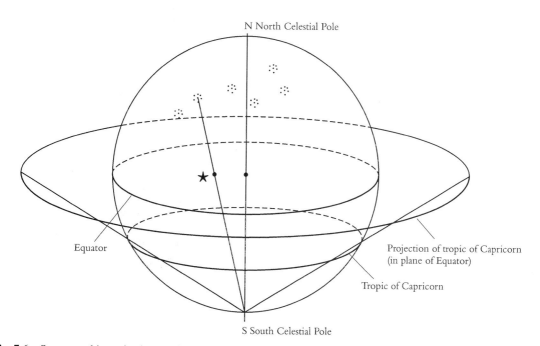

**Fig. 7.6** Stereographic projection used to make the star map for a planispheric astrolabe, with the South Celestial Pole as the centre of projection. This is the form of projection used on almost all medieval astrolabes.

of the tropics and the circle of the ecliptic (the annual path of the Sun) came out as circles in the star map of the astrolabe. It was thus relatively easy to draw them correctly.

As Ptolemy's text shows, stereographic projection was known in the second century AD, but it is not completely certain that Hellenistic astronomers also knew instruments like the majority of medieval astrolabes, properly called 'planispheric' astrolabes. When, in the *Almagest*, Ptolemy describes an instrument he calls 'astrolabion' ($\dot{\alpha}\sigma\tau\rho o\lambda\alpha\beta\iota o\nu$), it is an instrument rather like an armillary sphere, that is a model of the heavens showing the sphere by means of circles, some of the circles being adjustable so that they can be used for making observations. In the hope of avoiding confusion, this instrument, which was also used in some medieval observatories in the Islamic world, is now usually called an 'armillary astrolabe'. However, Ptolemy's treatise on astrology does refer to an instrument that probably was something like a planispheric astrolabe, and various references in later authors suggest very strongly that astrolabes of the common medieval type, with a flat star map, were known to astronomers at least from the late fourth century onwards.

In any case, Commandino, unlike Danti, does not seem to have had much interest in mathematical instruments as such. His reason for studying Ptolemy's *Planisphere* is that it deals with what Commandino, adopting the Greek term, calls 'scenography'. This subject, he says in the letter dedicating the book to his patron Cardinal Ranuccio Farnese (1530–1565, see Fig. 7.7), is one of the three branches of optics recognised in ancient times and is of particular use to the architect, who

does not propose to reproduce actual dimensions and proportions; but concerns himself with what will appropriately convey the appearance.

Such a programme means that

when he wishes to represent circles he sometimes does not draw circles but ellipses, and squares he makes longer on one side.

There are no references to the activity of painters, perhaps because Commandino wishes to confine himself to scenography in its narrow sense, namely the art of making stage scenery, which is the use of perspective for which he had the most secure evidence from ancient times. The only modern practitioner to whom Commandino refers by name is Vignola. He says Ranuccio Farnese's interest in Vignola's work, both as an architect (see Fig. 7.8) and as a scenographer, makes the Cardinal particularly appropriate as dedicatee of this edition of Ptolemy's work on scenography.

Commandino's insistence on the connection of the *Planisphere* with perspective may be partly due to his stated opinion that it is the only ancient work on whose behalf such a claim can be made. What he says is

Concerning what is to be done in such matters [i.e. drawing in perspective] we have nothing in writing from the ancients, that I know of, apart from these few words, in which Ptolemy deals with circles.

This opinion is perhaps a little curious, since Ptolemy's *Geography* (a text well known in the sixteenth century) does also partly deal with conical projection, as one way of making a flat map of the terrestrial sphere. Moreover, as we have seen, Egnazio Danti is of the opinion that the ancients have left us no writings on perspective. In any case, Commandino is undoubtedly correct in recognising that the stereographic projection Ptolemy describes in his *Planisphere* is formally related to the projection used in artificial perspective. Unlike sixteenth-century stage designers, however, Ptolemy does not propose to solve his problems by using string. His worked examples are geomet-

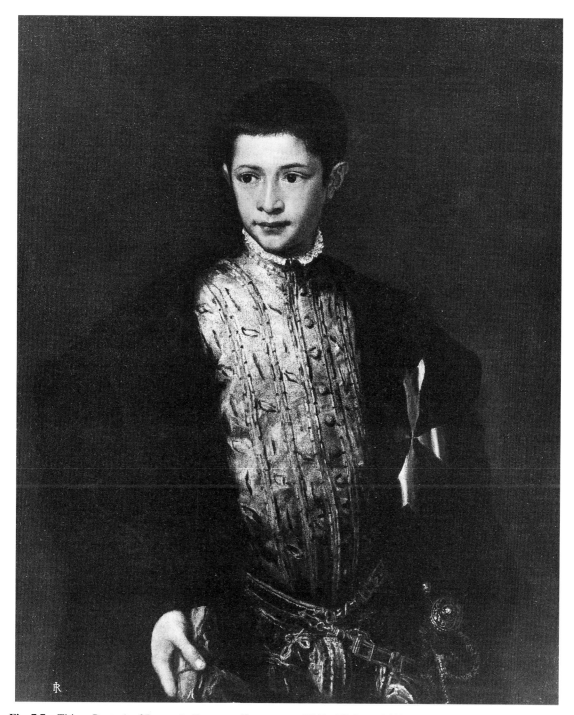

**Fig. 7.7** Titian, *Portrait of Ranuccio Farnese*, oil on canvas, 89.7 × 73.6 cm, 1542, National Gallery of Art, Washington, DC. Ranuccio, then aged twelve, is shown as a Knight of Malta.

Ptolemy does not propose to solve his problems by using string. His worked examples are geometrical in the manner of Euclid.

Unfortunately, worked examples are all we have. Moreover, as Commandino remarks, proofs have been omitted. (He provides the necessary proofs in his Commentary.) Furthermore, as it stands, Ptolemy's text is not complete. Since Commandino's time some additional fragments have been recovered, partly in Arabic and partly in the original Greek, but modern scholarship has little to add to Commandino's comment that

the book is wanting in Greek, and the book we have has been translated into Latin from Arabic in such a way that it is a very great undertaking to discover what the writer has in mind.

In fact, readers without an overwhelming preference for works of ancient authorship could find the mathematics of astrolabes better explained in several medieval texts, in particular the *Planisphere* of Jordanus de Nemore (*fl.c.* 1220), which is included in Commandino's volume.

**Fig. 7.8**   Villa Farnese at Caprarola, near Rome. The architect of the villa was Vignola.

It is possible that, in its original form, Ptolemy's *Planisphere* did have a general introduction. Commandino's Commentary, placed at the end of the volume, with a separate title page, is intended to provide it with one. In properly learned style (and, of course, in Latin) Commandino sets out to explain perspective in the most general terms possible. There are bizarre elements in this. As can be seen in Fig. 7.6, the projection used to obtain images of celestial circles has its equivalent of the picture plane, that is the plane of the celestial equator, cutting through the object to be represented, that is the celestial sphere. This does not affect the mathematics, but it produces a set-up unlike anything found in perspective treatises addressed to painters. Commandino draws attention to this fact by enumerating the possible different relationships of object and picture plane, though he does not actually treat the cases separately. Interestingly, in his letter to Ranuccio Farnese, he notes that in the absence of ancient sources he has made use of modern ones:

in our own time competent painters and architects have left descriptions of their actual practice in this matter, which have been very useful to me in following the reasoning of this book.

Commandino's work on the *Planisphere* was written in Rome, where he was then living, but he was a native of Urbino, and had worked there for many years, so it is highly probable that the modern works on perspective that he consulted included Piero della Francesca's *De prospectiva pingendi,* a copy of which was in the Ducal Library at Urbino.

There are, all the same, few direct traces of Piero's work to be found in Commandino's treatment of perspective. He begins by stating the problem in a standard way:

To draw the seen figure as it would appear in the proposed plane. Which is nothing other than to draw the common section of the proposed plane and the visual cones or pyramids, by which the figure is seen.

However, his actual treatment of the problem, which includes many diagrams and extends over 35 quarto pages, is very far from standard.

This may go some way to accounting for Daniele Barbaro's having found it, as he says, 'dark and difficult'. Commandino is presumably addressing himself to mathematicians, that is to professional mathematicians, the people who would be expected to take a serious interest in the details of the content of Ptolemy's far from elementary work.

Ptolemy is concerned only with circles. Commandino none the less begins with rectilinear figures. The first is a rectangle with one side lying along the ground line of the picture plane. The first of the series of accompanying diagrams is shown in Fig. 7.9. Commandino does not explain which cases are shown in his diagrams. This diagram refers to the simplest case, in which the set-up is symmetrical, that is when the plane through the eye perpendicular to the ground plane bisects the given rectangle. Further diagrams show asymmetrical cases. Commandino's proof applies to all cases. The symmetrical figure has been chosen merely because it is easiest to compare with those found in earlier treatments of similar problems.

Commandino first mentions the simplest case of all:

And if the eye is placed so as to be in the same plane as the figure that is seen, it [the figure] will appear as a single line, namely the line that is the common section [that is, the intersection] of the plane containing the figure and that of the picture.

It is possible that this is a reminiscence of the quasi-ground plans found in the work of Piero della Francesca, but it is a mathematician's way of dismissing them. We are then given instructions for drawing the first stage of the diagram shown in Fig. 7.9. The rectangle to be viewed is *abcd,* the

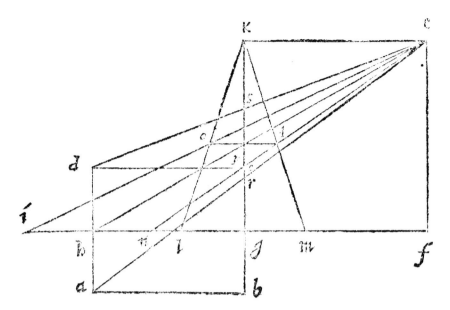

**Fig. 7.9** First in the series of figures for the summary of perspective in Commandino's commentary on Ptolemy's *Planisphere* (Venice, 1558, p. 2 *verso*).

eye is at *e*, and its height above the ground plane is *ef*. The line *fg* is constructed as perpendicular to *bc* and produced to cut *ad* in *h*. The line *gk* is constructed as perpendicular to *gh* and equal to *fe*. Thus far, apart from the definition of *ef*, we seem to be working in the plane, but, to establish that *gk* and *fe* are parallel, Commandino describes them not as both perpendicular to the same line (*gh*) but as both perpendicular to the same plane, giving a reference to *Elements*, Book 11, Proposition 6. The line *gh* obviously represents the ground plane but is also to some extent treated as a line in its own right, being called the *linea recta* of the plane (as *gk* is of the picture plane). This expression literally means 'straight line' or 'right line', and one might compare it with the term *latus rectum* used in connection with conic sections, but to avoid confusion it has been translated 'characteristic line'. The use of the name of the line to designate the plane in fact makes it difficult to be sure, at times, how many dimensions Commandino's diagram is meant to have. If we draw it as we go along, it will now look as in Fig. 7.10. Commandino's description is

Accordingly let the figure *abcd* be understood as lying in the plane whose characteristic line is *fg*: and the picture in a plane whose characteristic line is *gk*; so that the common section of the plane drawn through the lines *ef*, *fg* and the picture plane will be the line *gk*, and indeed the common section of the picture & the plane containing *abcd* will be the line *bc*.

The first case of the problem is then restated in precise terms:

Now the figure *abcd* is to be drawn in the picture plane *gk* in such a way as it appears to an eye placed at *e*; whose height from the plane is, as we said, *ef*; and, further, its distance from the picture is the line *fg*.

There follow instructions for constructing the points *l*, *m* (such that *gl* = *gb*, *gm* = *gc*) and *i*, *n* (such that *li* = *bn*, *mn* = *cd*). The points *l*, *m* are then joined to *k*, and *h*, *i*, *n* are joined to *e*, as shown in Fig. 7.11. In this figure the rectangle *abcd* has been omitted, since it makes no contribution to this

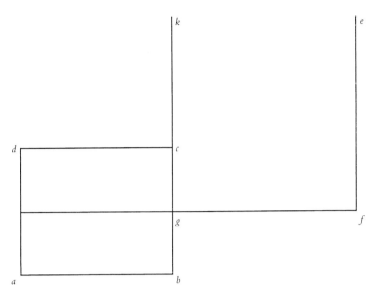

**Fig. 7.10** First stage of the diagram for Commandino's summary of perspective in his commentary on Ptolemy's *Planisphere* (1558), showing the set-up. The eye is at *e* and the picture plane is shown by *kg*.

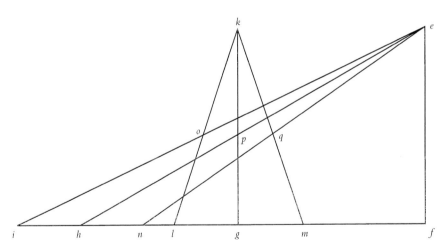

**Fig. 7.11** Second stage of the diagram for Commandino's summary of perspective in his commentary on Ptolemy's *Planisphere* (1558). The distances from *i*, *h*, *n*, *l*, and *m* to *h* have been given. It is to be proved that *o*, *p*, *q* are collinear and that the line *opq* is parallel to *lm*.

part of Commandino's work, in which he shows that *o*, *p*, *q* are collinear, and that the line *opq* is parallel to *lm*. This is a preliminary to the main result, which is that

the figure *abcd* will appear in the picture in the form *olmq*.

The proof that *o*, *p*, *q* are collinear depends upon considering the pairs of similar triangles *iol*, *koe*; *hpg*, *kpe*; *nqm*, *kqe*. These yield various sets of equal ratios which are then manipulated to give different sets which prove the similarity of the pairs of triangles such as *eop*, *eih*, thereby (together with a reference to *Elements*, Book 6, Proposition 2) establishing what Commandino wishes to prove.

We then return to three dimensions. The points *a*, *d* are joined to *e* and the intersection of the triangle *ead* with the picture plane is called *rs*. The line *rs*, on which the point *p* also lies, is parallel to *ad* (because *ad* is the line in which a plane parallel to the picture plane would intersect the triangle *ead*) and therefore, since *ad*, *bc* are parallel, the lines *rs*, *bc* are parallel.

The relevant part of Commandino's diagram (see Fig. 7.9) is shown in Fig. 7.12. The drawing conventions are such as to make Commandino's detailed verbal description more necessary than it may otherwise seem to a modern reader. Figure 7.13 shows the set-up in three dimensions in accordance with today's drawing conventions. We now consider the triangle *klm*, imagined as in the plane *ehf* (see Figs 7.9 and 7.14). This triangle is then applied to the picture plane so that *k* and *p* remain fixed. Commandino continues

Thus since the line *lm* will have been applied to the line *bc*: the line *oq* will also apply itself to the line *rs*: & they will be one and the same line; for the four points *oqrs* are in the same plane, which also contains *p*, and is parallel to the plane of the figure that is seen. So the point *o* falls in *r*, & *q* in *s*; since the line *po* is equal to the line *pr*: & *pq* to *ps*.

The final assertion is proved by considering the pairs of similar triangles *eha*, *epr*; *hpg*, *epk* and then manipulating ratios (*permutando*, *componendo* and so on) to obtain the desired equality. Commandino sums up with his customary care for precision rather than concision:

Therefore since the points *bc* are seen in the points *lm* of the figure that is drawn [i.e. the figure drawn in the plane *ehf*]: & the points *ad* in the points *oq*: the whole line *bc* will be seen in the whole line *lm*: and the line *ad* in the line *oq*: and on that account *ba* in *lo*: & *cd* in *mq*. Therefore the whole figure *abcd* will appear in the picture in that form in which we drew the figure *olmq*.

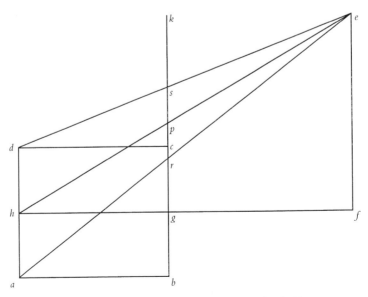

**Fig. 7.12** Part of the first diagram for Commandino's summary of perspective in his commentary on Ptolemy's *Planisphere* (1558). The part shown is that used at the beginning of the second part of Commandino's proof.

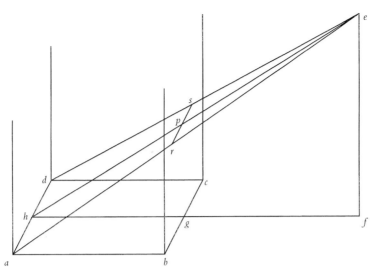

**Fig. 7.13** Modern version of part of the first diagram for Commandino's summary of perspective in his commentary on Ptolemy's *Planisphere* (1558). The part shown is that used at the beginning of the second part of Commandino's proof (cf. compare Fig. 7.12).

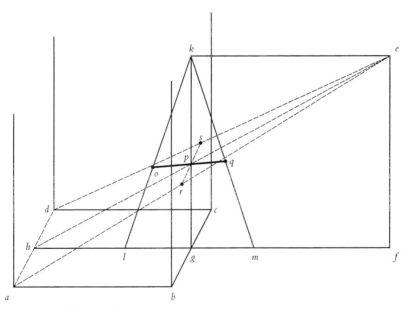

**Fig. 7.14** Modern version of the first diagram for Commandino's summary of perspective in his commentary on Ptolemy's *Planisphere* (1558) (cf. Fig. 7.9).

Nor does Commandino's professionalism let up at this point: the next paragraph begins 'ALITER'— 'alternatively'. The accompanying diagram is shown in Fig. 7.15, the first part to be used is the ground plan, shown separately in Fig. 7.16. (N.B. The points *i*, *n* in this new diagram are not the same as the points with the same letters in previous diagrams.) A figure is then drawn in the

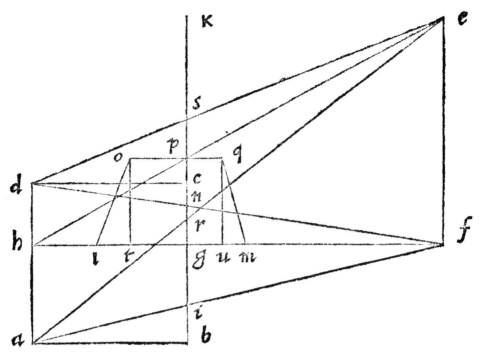

**Fig. 7.15** Diagram for Commandino's summary of perspective in his commentary on Ptolemy's *Planisphere* (1558). This diagram is that used for the second version of the proof whose first version referred to the diagram shown in Fig. 7.9. The points *i*, *n* in this new diagram are not the same as the points with the same letters in previous diagrams.

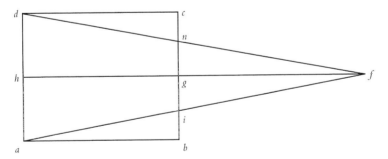

**Fig. 7.16** Copy of part of a diagram for Commandino's summary of perspective in his commentary on Ptolemy's *Planisphere* (1558). It shows the ground plan from the diagram in Fig. 7.15.

plane *ehf* in rather the same way as before, constructing points *l*, *m*, *t*, *u* so that their distances from *g* are the same as those of *b*, *c*, *i*, *n*. The point *p* is, as before, the intersection of the line *he* with the vertical through *g*. The points *o* and *q* are found as the points of intersection of verticals through *t*, *u* and the horizontal through *p*. Next, the trapezium *olmq* is rotated about *gp* so as to come into the plane of the picture and the proof proceeds much as before, with appropriate references to Euclid throughout. That is, the 'aliter' proof works in much the same way as its predecessor.

These two proofs have included proofs of results that Piero della Francesca took for granted by his use of quasi-ground plans, and have dropped explicit references to Euclidean optics. Moreover, the fact that what Piero made a series of propositions has now become a single one is a further indication that Commandino is turning his back on the practical instructional tradition. He has given us mathematics for mathematicians.

What is missing from his apparently thorough mathematical treatment is a proof of the convergence of orthogonals. As we have seen, Piero did all but provide such a proof (in Book 1, Proposition 8; see Chapter 5) and the convergence is visible in many of his diagrams. Commandino comes less close to giving a proof. In principle, since the width $bc$ ( $= ad$) of the rectangle is arbitrary, it follows that the images of lines parallel to $ba$, $cd$ will always meet at $k$. However, Commandino's exposition is so laboriously explicit that one cannot help suspecting that if he had thought about this proof of convergence he would have mentioned it. The omission may be explained by the fact that the convergence of orthogonals is not relevant to Ptolemy's work. On the other hand, it was the element in perspective construction with which architects and other practitioners seem to have been most concerned. Quite apart from considerations of the mathematical difficulty of Commandino's treatment, it is not hard to see why, in writing about perspective for his fellow patricians, Daniele Barbaro chose to base himself upon Piero della Francesca's work rather than Commandino's.

The remainder of Commandino's summary of perspective is concerned with horizontal rectangles lying in the ground plane whose nearest sides are not parallel to the ground line or are set back from it, and eventually with an irregular quadrilateral. Like Piero della Francesca (see Fig. 5.12), Commandino deals with these cases by constructing perpendiculars to the ground line from the vertices of the figures concerned.

Since Ptolemy deals with circles, Commandino clearly has to do so too. The circle appears as the circumcircle of a regular octagon, and discussion then becomes directed towards the properties of the cone with its apex at $e$. This part of the work is beyond the scope of all previous work on perspective, and (with hindsight) belongs more properly in a consideration of the history of the study of conic sections. We shall accordingly postpone discussion of it until the next chapter. After the first appearance of the circle, Commandino reverts to more conventional problems, considering surfaces lying above the ground plane, and at the very end of his discussion of perspective indicates very briefly how the methods he has described can be used to obtain the images of a regular tetrahedron and a cube. In the final sentence, he mentions that it can also be applied to any other body. Intellectually, we are as far as possible from the series of worked examples that characterise treatises on perspective addressed to artists.

## Explaining practice: Giovanni Battista Benedetti

As we have already mentioned, Commandino's work seemed 'dark and difficult' to Daniele Barbaro. Moreover, it seems not to have been very interesting to Egnazio Danti, whose total account of it is:

Furthermore, Commandino gave geometrical demonstrations of how an object appears to the eye in Perspective in all cases which can present themselves.

The emphasis upon Commandino's provision of proofs is certainly reasonable, but Danti's summary makes the work seem far more comprehensive than it actually is. The implication is that Commandino deals in a general manner with three-dimensional figures. He does not. Danti's comment in fact seems little more than a reflection of Commandino's high reputation. And it suggests that Danti had not made a detailed study of the text in question. Perhaps one good reason for his neglect may have been the fact that Commandino's text would not have been useful to the readers Danti was addressing in his edition of Vignola's perspective treatise. Commandino's work is relatively difficult, as mathematics, but the problems it treats are only the simplest of those with which a practitioner would be concerned. For instance, Commandino explains the mathematics of one possible method of constructing a square but does not tell one how to turn the square into a *pavimento*. Cases have been chosen for their mathematical interest, not for their usefulness to the practitioner.

On the other hand, when we look at the actual mathematics of Commandino's work, what we find is thoroughly conventional. Almost everything is dealt with, as far as possible, in two dimensions (and using similar triangles). Indeed, the usual drawing conventions, in the superposition of figures from perpendicular planes, seem to have contributed to the overall shape of Commandino's proofs. In giving proofs, and in his choice of cases, Commandino is outside the usual conventions of writings on perspective, but the style of his mathematics is completely in line with the normal conventions of other mathematical texts.

The situation is almost entirely the other way about when we come to the work of Giovanni Battista Benedetti (1530–1590). The perspective problems he deals with are apparently dictated by the normal practice of painters, architects and scenographers, but the manner in which they are treated is entirely new. Benedetti jumps straight into three dimensions. His treatment resembles Commandino's only in that both are clearly addressing themselves exclusively to competent mathematicians.

Unlike Commandino's, at least some of Benedetti's mathematical origins are connected with the practical tradition: he was, he tells us, a pupil of Niccolò Tartaglia. He then adds, in the standard sixteenth-century Michelangelesque manner, that of course he learned very little from him. This may be true in the sense that Tartaglia's most lasting achievements were in algebra whereas Benedetti was by nature a geometer. Their area of intellectual overlap was in their studies of the motion of heavy bodies. We have already mentioned Tartaglia's 'new science' of gunnery. Benedetti's work is directed to more abstract issues and is notable for describing some thought experiments concerning free fall which (with hindsight) are relevant to later, and much better known, work on the same subject by Galileo Galilei. Galileo does not refer to Benedetti's work, which in the circumstances is certainly not very strong evidence that he did not know it—but that is part of a different story.

Benedetti was born in Venice, though his family may have had connections with Spain. He writes in Latin, in a rather intricate style that suggests a university education, as does the nature of some of the philosophical topics he discusses, but we do not know that he did attend a university. In fact, we know very little about Benedetti's personal life, except that he appears to have been sufficiently rich not to depend upon his patrons for his livelihood. All the same, he seems largely to have lived the life of a courtier. In the early 1560s he was at the court of Parma, and the remainder of his life was spent at the court of the Grand Dukes of Savoy in Turin. There, he was (of course) requested to give his opinion on calendar reform, but was also asked about more elementary matters, as he tells us near the beginning of his treatise on arithmetic (*Arithmetical theorems*, *Theoremata arithmetica*):

The Most Serene Duke of Savoy asked me by what reason it can be known scientifically and theoretically (as one says) that the product of two fractional numbers is less than either of the factors that produce it. To which I replied that in one's mind and thought one must imagine the fractional factors and the fractional product not as being of one and the same nature but as very different.

That is, in answering this question, Benedetti drew attention to the essentially geometrical problem of dimensionality. In accordance with the Euclidean tradition, the numbers will be represented by lengths and the product will be an area. The treatise that follows on from this first theorem contains a large number of diagrams, as aids to visualising what is going on. Even in writing about arithmetic Benedetti is thinking like a geometer.

Benedetti's short treatise on perspective was published, with the treatise on arithmetic, a number of other works (all short), and a large number of letters on scientific topics, in a volume entitled *A book containing various studies of mathematics and physics* (*Diversarum speculationum mathematicarum et physicarum liber*, Turin, 1585). Like the treatise on arithmetic, the work on perspective may have originated from a query, for after its splendidly brisk beginning it proceeds to deal with just the sort of question that a baffled amateur might have asked of a professional. There is, however, no hint of such an origin in the way Benedetti's treatise begins. He says

Since no-one so far (that I know) has given a perfectly correct account of the reasons underlying the operation of perspective construction, I thought it worthwhile to give some discussion of them.

He then proceeds to consider an error (for whose actual occurrence we have plenty of other evidence, some of it pictorial):

For many of those who lay down rules for this kind of operation, in their ignorance of the true reasons make various and different kinds of errors, as, for example, when in figure A below, wishing to degrade (as they call it) the rectangle .q.a. in the triangle .i.d.q., they draw a line parallel to .q.d. from the point *B*, the point of intersection of the line .o.l. with the side .i.d. of the triangle, & (in the same case) equally draw the parallel line from the point *Z*, the point of intersection of the same line .o.l. with the perpendicular line .x.i. …

Benedetti's figure A is shown in our Fig. 7.17. Like all the illustrations in his book, it is adequately large but not in the least elegant. It does, however, show a touch of humour: the rectangle has been called *quad*. Unfortunately, this piece of light relief gets lost in what follows since Benedetti adopts the standard mathematical practice of referring to quadrangular figures by the letters marking the ends of one diagonal, thus turning *quad* into *qa*.

The reader is clearly expected to be familiar with this type of perspective construction, and to recognise, for instance, that the point *o* represents the eye, the point *p* the 'foot of the eye' (the foot of the perpendicular from the eye to the ground plane), and that the point *l* has been constructed so that $ld = da$. (The existence of two points called *i* is explained later.) The error of which Benedetti is complaining is evidently the result of a confusion between the distance point construction, in which the viewing distance of the picture would be the longer version of *oi* and it would be correct to draw the parallel through *B* (cf. Figs 2.8 and 6.6), and the Albertian construction, in which the viewing distance would be the shorter *oi* and the parallel should be drawn through *Z* (cf. Figs 2.4 and 5.7). To decide the matter of which construction is mathematically correct, Benedetti considers the actual 'pyramid of vision'. The accompanying diagram is three dimensional (see Fig. 7.18). This diagram is like those found as illustrations in works on natural optics, but Benedetti's pyramid is much more complicated than any of its predecessors.

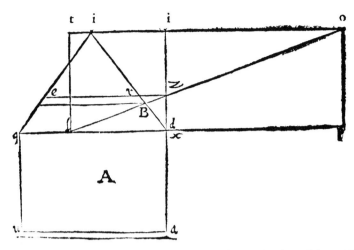

**Fig. 7.17** Giovanni Battista Benedetti, *A book containing various studies* ... (Turin, 1585), p. 119. Figure A, showing constructions to find the perspective image of a rectangle *quad* ('to degrade *quad* in the triangle *idq*').

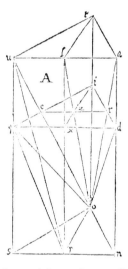

**Fig. 7.18** Giovanni Battista Benedetti, *A book containing various studies* ... (Turin, 1585), p. 120. Figure A, in three dimensions, that is 'solid' (*corporea*), corresponding to the two-dimensional diagram ('surface', *superficialis*) shown in our Fig. 7.17.

In Benedetti's figure, the observer is shown (supine) at the bottom of the diagram, his eye at *o* and his foot at *p*. This may be a comment on the prevalence of illusionistic ceiling paintings, but it is more probable that the unnatural orientation was chosen because it allows an undistorted view of the ground plane *ausn*. From this figure, Benedetti has no difficulty in showing through which point the parallel should be drawn in the two-dimensional version (shown in our Fig. 7.17). The sophisticated part of his mathematics is to establish the properties of the three-dimensional figure itself. Readers who think it is 'obvious' that we have a right triangular prism and so on, are not the kind of mathematician that Benedetti has in mind. Here one's powers of visualising things in three dimensions conflict with Euclid's style, in *Elements* 11, that of proceeding by the rigorous method of deductive proof.

As a good geometer, Benedetti is on Euclid's side. We start with the (horizontal) ground plane, in which we are given the rectangle *quad*, the fact that *qd* lies along the ground line of the picture, and that *x*, the foot of the perpendicular from *p* to *qd*, lies at the mid-point of *qd* (which will turn out not to matter). We are also given that the line *op* and the picture plane are vertical. The rest Benedetti proves, with numerous references to *Elements* 11. That is, he establishes that we have a right triangular prism: with the three planes *tua*, *iqd*, *osn* parallel to one another and perpendicular to the ground plane *uasn* and the planes *utos*, *aton* and *lop*. Comparing the figure *lop* in the pair of diagrams *A* then establishes that since the viewing distance is *px* the point *Z* (shown in both figures) gives the correct height for the parallel.

What Benedetti says about the use of *B* instead of *Z* is very brief and does not refer to it as having any connection with an alternative construction. Instead he says, after a reference to Witelo's treatise on optics, that it would be correct to use *B* only if the angles *iqd* and *idq* were made larger. As it stands, the significance of this is far from clear. Moreover, Witelo's treatise, which is based on the work of Ibn al-Haytham, is far from elementary, though in the sixteenth century it did displace the simpler Euclidean optical texts as the standard reference for optics (as far as professional mathematicians were concerned). It is thus not altogether surprising that Pietro Accolti, reading what Benedetti has to say (and giving an exact reference, thus leaving us in no doubt about what he has been reading), should have been misled into believing that Benedetti had proved that the Albertian construction was the *only* correct one, and hence, in 1625, dubbing it 'the legitimate construction' (*costruzione legittima*). Though clearly not a very able mathematician, Accolti was an honoured member of the illustrious Florentine Accademia del Disegno, and a distinguished writer on artistic and other subjects, so this silly name for the construction was widely accepted. Since, as we shall see, Benedetti goes on to consider a number of more general cases, he does in effect prove that the distance point construction is also mathematically correct. Accolti does not refer to the later parts of the treatise.

Having established the correspondences between the 'surface' (*superficialis*) and 'solid' (*corporea*) versions of his figure A, Benedetti then proceeds to show that analogous correspondences will hold in the case shown in the pair of figures B (see our Fig. 7.19), in which *x*, the foot of the perpendicular from *p* to the line of intersection of the picture plane with the ground plane (the line *qd* produced), lies outside the segment *qd*. If *qd* is thought of as being the actual ground line of the picture, then this corresponds to the centric point of the perspective lying outside the picture—a case Piero della Francesca had explicitly excluded from consideration (see Chapter 5). It was, however, of relevance for designers of stage scenery (see below). The chapter dealing with the case shown in Benedetti's figures B is extremely brief

since, when the figures A have been understood, it will also be easy to understand the two subsequent ones, B.

This is, of course, true, but serves to remind us that we are not dealing with a 'practical' text in which exposition proceeds through series of worked examples.

In the case shown in his figures B, Benedetti has the point *Z* on a vertical through the centric point *i*, both in the surface and in the solid figure, whereas in A it was on this vertical only in the solid one (hence there being two points called *i* in the surface figure A). In his next chapter, over

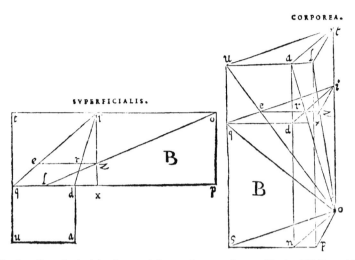

**Fig. 7.19**  Giovanni Battista Benedetti, *A book containing various studies …* (Turin, 1585), p. 122. Figures B, in two dimensions ('surface', *superficialis*) and in three dimensions ('solid', *corporea*). Compare with Figures A, our Figs 7.17, 7.18.

which the modern phrase 'without loss of generality' seems to hover, Benedetti shows that *Z*, the point giving the height of the parallel, can always be put on the vertical through the centric point *i*. The model general case is shown in his figure C (see our Fig. 7.20). The next chapter concerns a different form of generalisation: dropping the condition that the picture plane shall be perpendicular to the line of sight. Benedetti's pair of diagrams D is shown in our Fig. 7.21. It seems very probable that this new generalisation, like the generalisation of allowing the vertical line through the centric point to lie outside the triangle *iqd*, as shown in figures B (our Fig. 7.19), arises from considering the perspective schemes of side pieces in stage scenery, in the kind of set-up shown in

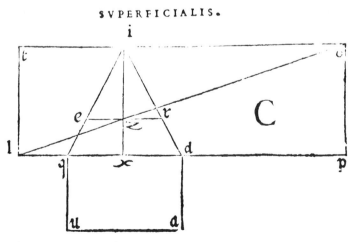

**Fig. 7.20**  Giovanni Battista Benedetti, *A book containing various studies …* (Turin, 1585), p. 122. Figure C, in two dimensions ('surface', *superficialis*), showing *Z* constructed on the vertical through the centric point *i*. Benedetti shows this case to be a general one.

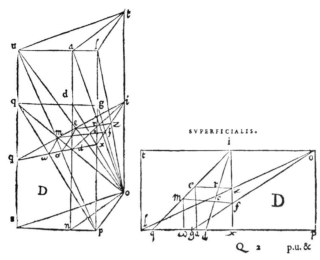

**Fig. 7.21** Giovanni Battista Benedetti, *A book containing various studies …* (Turin, 1585), p. 123. Figure D, in two dimensions ('surface', *superficialis*), and in three ('solid', *corporea*), showing the case where the picture plane *iqd* is not perpendicular to the line of sight *oi*.

our Figs 6.19 and 6.20. For these side pieces, we have a combination of the conditions of B and D, that is, the centric point lies outside the picture field and the picture is seen at an angle. As Danti's diagram for Vignola's treatise (our Fig. 6.19) shows, in practice the solution to the problem involved the use of string. Benedetti's discussion is evidently addressed to mathematicians, not to practitioners, though their activity has apparently contributed to his choice of cases to discuss.

The construction shown in Benedetti's figure C provides the basis for finding the perspective image of a general point in the ground plane. The construction is shown in the pair of diagrams E, one surface and one solid (see our Fig. 7.22). Benedetti has used the same letters for the points *l*, *m* in their original positions below *qd* and for the points corresponding to them on *qd* (or *qd* produced) that are used to find the perspective images of the lines *ua* and *bm*. The problem of finding the image of *bm* is formally the same as that of finding the image of *ua*, which no doubt explains why Benedetti has used equivalent constructions, involving joining *l*, *m* to *o*. This method brings out the mathematical equivalence, but is not a good method in practical terms, since it involves drawing more lines that may lie at least partly beyond the edge of the picture field. In fact Benedetti seems generally unconcerned with the picture field, apparently preferring to think in terms of the picture plane, defined by the triangle *iqd*.

None the less, the problems he treats continue to have the air of having been suggested by the activity of practitioners. One of these problems, which proves to be of some mathematical interest in its own right, seems to have been inspired by one of the mechanical drawing aids recommended by Dürer and others. The instrument in question is the frame with crossing adjustable strings shown in Figs 6.7 and 6.8. The figure for Benedetti's discussion is shown in our Fig. 7.23. In this case, Benedetti provides only the surface figure, and no lines are drawn beyond the edges of the triangle *iqd*. The habit of using woodcuts for printed illustrations had lowered the aesthetic standard of diagrams in mathematical texts—the effect is noticeable even in manuscripts—but apart from this Benedetti's diagram K has a fifteenth-century look to it. The statement of the problem does not mention any drawing apparatus, but is in exclusively mathematical terms:

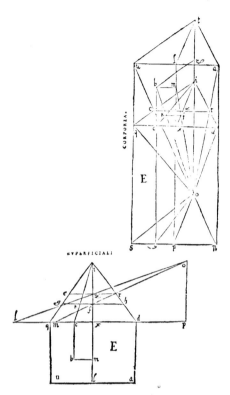

**Fig. 7.22** Giovanni Battista Benedetti, *A book containing various studies* … (Turin, 1585), p. 125. Figure E, in two dimensions ('surface', *superficialis*), and in three ('solid', *corporea*), showing the construction of the perspective image of a general point *b* in the ground plane.

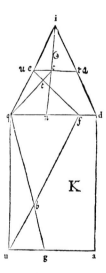

**Fig. 7.23** Giovanni Battista Benedetti, *A book containing various studies* … (Turin, 1585), p. 120. Figure K (first version), in two dimensions only, showing the construction of the perspective image of a general point *b* in the ground plane when the position of the point is defined by the intersection of two diagonal lines (*uf* and *qg* in the ground plane and their images, *ef* and *qc*, in the picture).

So let there be in the figure K a point *.b.* in the perfect parallelogram [*quad*] whose position is to be found in the degraded one *.e.q.d.r.*

The solution is appropriately simple:

Now from any two of the four corners *q.u.a.d.* there are drawn two line segments *.q.g.* and *.u.f.* through the point *.b.* as far as the sides *.q.d.* and *u.a.* [of the rectangle *quad*] and so that their ends *.g.* and *.f.* will lie within the sides *.q.d.* and *.u.a.*, that is so that they do not cut the two sides *.q.u.* or *.d.a.* Then let the point *.f.* between *.q.* and *.d.* be joined to the degraded corner *.e.*, which corresponds to *.u.* in the perfect figure, by means of the line *.e.f.*, which will be *.u.f.* degraded in our plane. Then let there be taken the point *.n.* on the line *.q.d.*, such that its distance from *.q.* is the same as that of *.g.* from *.u.*, and let there be drawn the line *.i.n.* and let it cut the line *.e.r.* in the point *.c.*, which, from what we have said above, corresponds to the point *.g.* Next, drawing the line segment *.q.c.* it will be seen to correspond to the line *.q.g.* which, cutting the line *e.f.* in the point *.t.*, obviously corresponds to the point *.b.* as all will recognise.

After this very simple example, Benedetti turns to the case where *b* lies on one of the sides of the perfect figure, say *qu*. His diagram is shown in our Fig. 7.24. In terms of the drawing apparatus illustrated by Dürer and Danti, this case is of no interest at all. Moreover, it is clearly a simple version of the general case shown in Benedetti's figure E: if *b* lies on *qu* then the line *ci* coincides with the line *qi*. Benedetti is presumably dealing with the problem in this new way purely for its mathematical interest, and for the sake of mathematical completeness. What it gets him is a partial

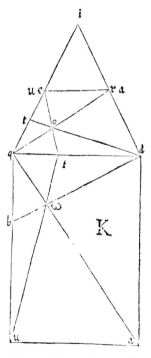

**Fig. 7.24**  Giovanni Battista Benedetti, *A book containing various studies* … (Turin, 1585), p. 130. Figure K (second version), in two dimensions only, showing the construction of the perspective image of a point *b* lying on one side of the rectangle *quad*, when the position of the point is defined by the intersection of diagonal lines (*qd*, *uf* and *fω* in the ground plane and their images, *qr*, *ef* and *do*, in the picture).

converse of Desargues' famous theorem on triangles in perspective. This theorem was discovered some time between 1639 and 1648 (see Chapter 8). It is now usually stated in the form:

*If two triangles are in perspective, then the meets of corresponding sides are collinear.*

In Benedetti's example, the triangles in question are *uωb* and *eot*, and their being in perspective (with centre of perspective at the observer's eye) is much clearer in the three-dimensional diagram, which Benedetti has not supplied for either version of K. The omission may reflect a feeling that he is now concerned not with the fundamentals of perspective but merely with some technical results. Since in his figure K Benedetti has used the letter *o* for a point in the picture plane, our substitute for the missing three-dimensional diagram (Fig. 7.25) has called the eye *Y*. The orientation of the diagram in Fig. 7.25 is different from that of Benedetti's own solid diagrams so as to make it look more like those usually drawn to illustrate Desargues' work. (The comparison is, of course, essentially anachronistic, so it is probably just as well that this fact should be visible in the diagram.) What Benedetti has proved is that since the points *d*, *q*, *f* are collinear the triangles *uωb*, *eot* are in perspective. Desargues' theorem in this case would be that since the triangles *uωb*, *eot* are in perspective the meets of corresponding sides, that is *d* (the meet of *bω* and *to*), *q* (the meet of *ub* and *et*) and *f* (the meet of *uω* and *eo*), are collinear. Desargues' theorem holds for all triangles, whether or not they are in the same plane. Benedetti has proved his result (without stating it) only for triangles that are not coplanar, so what he has is a partial converse of the theorem.

Benedetti evidently does not recognise that he has proved this result, or, at least, does not see it as having an independent interest as a *general* result, that is as a theorem. One has the slightly

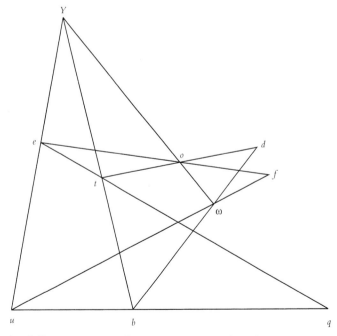

**Fig. 7.25** Three-dimensional diagram corresponding to the second version of Benedetti's Figure K, to show its connection with Desargues' theorem. Benedetti does not supply an equivalent to this figure. The eye, the centre of perspective, is at *Y*, the two triangles in perspective are *uωb* and *eot*, pairs of corresponding sides meet at the points *d*, *q* and *f*, which are collinear.

absurd situation of historians, equipped with 20/20 hindsight, shouting 'look behind you!' like children at a pantomime. This is fun, of course, but in the present instance it is also of some serious interest because it illustrates both the rigour of Benedetti's work and the nature of its limitations. Benedetti has set out to explain some well known parts of the practice of perspective, that is to explain them in mathematical terms in the rigorous deductive manner of what we may call the 'high' mathematical tradition, exemplified by Greek geometry. The activities of the practitioners have become sufficiently important to attract the attention of a first rate learned geometer. The practitioners themselves were probably a lot less interested in this than Molière's hero M. Jourdain was to learn (about 80 years later) that he had been speaking prose all his life. Benedetti's diagram D is not a viable alternative to using string in constructing the perspective of stage sets. On the one hand, Benedetti's work shows us that the activities of practitioners could direct the attention of mathematicians to new and fruitful areas of investigation: his three-dimensional diagrams explicitly introduce the centre of projection and therefore clearly bring us closer to seeing perspective as an example of projective geometry (that is to the work of Desargues). On the other hand, however, Benedetti's example also shows us how his work is limited by the concern with practical problems that apparently set it in train. He deals with the problems concerned, but shows no inclination to explore any further possibilities that the method of solution may itself suggest. He is behaving as what would now be called an 'applied' mathematician, giving original solutions to particular problems, but not, as a 'pure' mathematician would, investigating possible general implications of his methods. We shall find the same pattern in Benedetti's work on conic sections. In that, as in his work on perspective, we have what turn out to be first class original contributions to the high tradition of mathematics arising out of problems conceived within the practical one.

## The view from point *A*: Guidobaldo del Monte

Benedetti's treatise on perspective is a contribution to the tradition of learned mathematics in the sense that it undoubtedly marks an increase in understanding of the subject with which it is concerned. It is interesting, historically, to know that someone could think like that in the 1580s. Presumably, since Accolti came to read it, the work was known among experts on the theory of art, if not among artists themselves. However, Benedetti left behind no pupils eager to develop his ideas, and it would appear that, apart from confusing Accolti, the treatise had no effect at all upon later developments.

Commandino's work on perspective fared considerably better. There is no evidence it was more widely read than Benedetti's—that is, we do not find later writers quoting it—but the ideas it contained were developed, some of them at considerable length, by one of Commandino's ablest pupils, Guidobaldo del Monte. Guidobaldo was a nobleman, though the family had only been ennobled in the previous generation. All the same, Guidobaldo bore the title Marchese (Marquis) of Pesaro, dedicated his books not to patrons but to members of his own family, and made use of his social position to help Galileo Galilei obtain the post of Professor of Mathematics at the University of Padua in 1592. He is thus well connected intellectually as well as socially. Guidobaldo certainly took an interest in practical matters, for instance as Surveyor of Fortifications in Tuscany, and as author of the short book about a new kind of astrolabe to which we referred briefly in Chapter 5 in connection with its discussion of the drawing of ellipses (*The theory of universal planispheres*, *Planisphæriorum universalium theorica*, Pesaro 1579). None the less, one cannot help suspecting that his interest is to some extent that of the patrician rather than the practitioner. As a mathematician, he seems to have generally been competent rather than original. The new type of astrolabe he described was not of his own invention, but his interest in astronomy did extend to his doing some work on the reform of the calendar.

Almost everything in Guidobaldo's *Six books on perspective* (*Perspectivæ libri sex*, Pesaro, 1600) can be seen as a development of Commandino's concise comments on similar topics in his Commentary on Ptolemy's *Planisphere*. There is, however, one very important exception. Unlike either Commandino or Benedetti, Guidobaldo appears to be interested in the convergence of the orthogonals. His treatise begins with a very detailed treatment of the perspective images of lines. This leads him to the important result that not only orthogonals but any set of lines that are parallel in what earlier writers called their 'perfect' or 'proper' forms, will appear in the picture as a set of lines meeting at a point (unless they happen also to be parallel to the ground line of the picture). The build-up to this result is, however, extremely slow. Perhaps in reaction to Commandino's having lumped so much into his two initial propositions, Guidobaldo seems determined to divide his results as finely as possible. The rule seems to be that one case makes one proposition. So considering lines in various series of orientations divides each general theorem into several sub-theorems, each proved by essentially the same method—except that Guidobaldo seemingly cannot bring himself to say so, and prefers to repeat proofs line by line. Thus, though the content is far from elementary, and there are numerous references to *Elements* 11, the style has something of the didactic quality of abacus books.

The series of propositions concerning the images of sets of parallel lines starts with the theorem in Book 1, Proposition 24:

If the eye sees parallel lines, and the section is parallel to the parallel lines; the lines appearing in the section will be parallel to one another.

Guidobaldo's use of the word 'section' to mean the picture plane, which Commandino had called the 'picture', is presumably derived from Euclid's habit (copied by Commandino and Benedetti) of using the phrase 'common section' for what we should now call the 'intersection' of, say, a pyramid and a plane. The dropping of the word 'picture' probably also indicates that Guidobaldo recognises that what he has to say has rather little to do with the making of pictures, though his propositions all start with a reference to the eye. His name for the horizontal plane through the foot of the observer and the lower edge of the picture, the plane now usually called the 'ground plane', is the 'subject plane', and its intersection with the picture (the 'ground line' of the picture) is called the 'line of section'. (Slight variations on these terms were to become standard in the treatises of the following centuries.)

One corollary and three theorems after the passage just quoted, we have the first appearance of the point of concurrence, though it is not as yet given a name:

If the eye sees any number of parallel lines in the subject plane, which are not parallel to the line of section; and the section is perpendicular to the subject plane; the lines appearing in the section will all be concurrent to one and the same point, whose height above the subject plane will be the same as that of the eye.

The next theorem has the section no longer at right angles to the subject plane; in the next, some of the parallel lines are not in the subject plane; next all the lines lie in any plane passing through the line of section … and so on. In each case we have a point of concurrence, and Guidobaldo indicates its relation to the line of section and to the point $A$, the position of the eye. After Proposition 32, we are at last given a title for the point at which the images of sets of parallel lines meet: Guidobaldo names it *punctum concursus* ('point of concurrence'), and notes that it will be given the letter $X$ in his diagrams.

Two more corollaries, one more theorem and then another corollary follow before we come to the important general result about points of concurrence, which, since all the theorems are special cases, naturally takes the form of another corollary:

From this it is also obvious that all the parallel lines lying in the subject plane, & others, not lying in the subject plane, parallel to them, have a point of concurrence on the line parallel to the line of section and at a distance from it equal to that of the eye above the subject plane.

Curiously, in his enunciation of this corollary, Guidobaldo has omitted to refer to the exception provided by the case in which the set of parallels is parallel to the line of section. The brief proof of this corollary, like the much lengthier proofs of the results that have led up to it (the theorems and corollaries in Propositions 24 to 33) depends almost entirely on references to Euclid. Guidobaldo, who is writing in Latin, undoubtedly sees himself as making a contribution to the learned tradition of geometry. A diagram showing points of concurrence for the images of four pairs of parallel lines appears on the title-page of his treatise (see Fig. 7.26).

Although his work is written in the learned style, it seems likely that Guidobaldo's interest in points of concurrence arises from questions associated with artistic practice. We know that in the sixteenth century disputes sometimes arose about how many such points there should be in a picture, and on occasion there even seem to have been suggestions that the artist should be asked to put in more. One can easily imagine a mathematician being irritated by such muddled and unanswerable questions, perhaps even sufficiently irritated to give the matter some serious thought. In any case, the outcome of Guidobaldo's thoughts about points of concurrence does seem to have been considered useful by later authors, particularly mathematicians, who wrote about perspective for artists. In fact, Guidobaldo himself goes on to discuss quasi-practical problems. Most of his second book is taken up with problems of how to find the positions of the points of concurrence for various sets of parallels. The first proposition is the problem

Given the eye, and given parallel lines lying in the subject plane which are not parallel to the line of section, to find the point of concurrence [of the images of these lines] in a section that is at right angles to the subject plane.

Guidobaldo's diagram for this problem is shown in Fig. 7.27. Like all his diagrams, it is drawn in more or less naturalistic perspective (the parallels in the subject plane tend to look as if they were orthogonals even when the text tells one they are not). Readers are being invited to think in three-dimensional terms.

Clearly, solving this kind of problem about points of concurrence could deal with claims that an artists had put them in the wrong places. Similarly quasi-practical problems occur throughout Guidobaldo's treatise, and as a set piece finale we find a discussion of stage scenery. There is little interest here in attaching string to a peg labelled $A$ representing the eye, as in the text accompanying Danti's diagram of the practical solution shown in our Fig. 6.19. Guidobaldo is concerned with the mathematical point $A$, and the view from it. As a Marchese, he presumably had a higher chance than most mathematical writers of actually sitting at $A$. His treatment of the perspective of stage sets is unremittingly mathematical—and guaranteed to confirm any practitioner's preference

**Fig. 7.26** Guidobaldo del Monte, *Six books on perspective* (Pesaro, 1600), title-page, showing points of concurrence for the images of four sets of parallel lines.

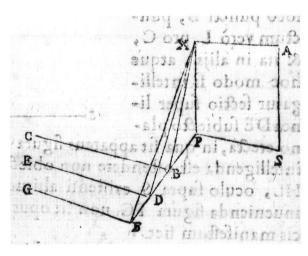

**Fig. 7.27** Guidobaldo del Monte, *Six books on perspective* (Pesaro, 1600), p. 54, diagram to illustrate Book 2, Proposition 1. The eye is at *A*, and we are given three lines *BC*, *DE*, *FG* in the subject plane, parallel to one another but not parallel to the line of section (*FP*). Their images meet at the point *X*.

for using string. Some of the accompanying diagrams are shown in our Figs 7.28, 7.29 and 7.30. Illustrations like these continue to appear in later treatises on stagecraft, apparently as a sort of fairy-tale ideal to which one may claim to aspire. Practical scenographers had, on the whole, little interest in publishing the tricks of their trade. They were being paid to provide surprises.

The relatively wide appeal of Guidobaldo's *Six books on perspective*, and its use by later authors, shows how far the general level of mathematical education had risen. References to Euclid's work on solid geometry clearly no longer looked intimidating. Actually, *Elements* 11 is not very difficult for twentieth-century readers. What seems to have been found difficult in the Renaissance is that it

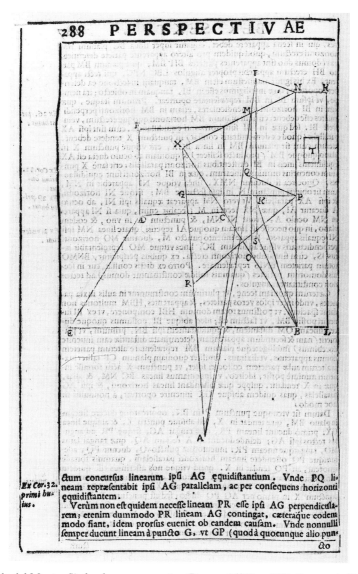

**Fig. 7.28** Guidobaldo del Monte, *Six books on perspective* (Pesaro, 1600), p. 288, diagram to illustrate Book 6, 'On stage scenery', showing central panel and two side pieces. The eye is at *A*. Compare this with Danti's diagram of a similar stage set shown in Fig. 6.19.

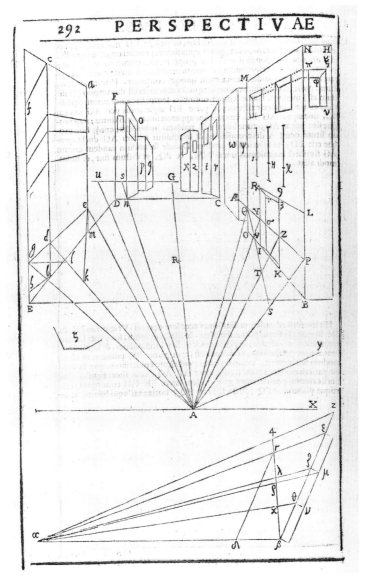

**Fig. 7.29** Guidobaldo del Monte, *Six books on perspective* (Pesaro, 1600), p. 293, diagram to illustrate Book 6, 'On stage scenery', showing central panel and side pieces, with an additional street indicated on the left. The eye is at *A*. Compare this with Danti's diagram of a similar stage set shown in Fig. 6.19, and the scenery of the Teatro Olimpico, shown in Figs 6.22 and 6.24.

requires one to think about geometry in three dimensions. The practical tradition, exemplified by the activities of surveyors and military engineers, seems to have familiarised mathematicians with essentially three-dimensional problems. The work of Benedetti and Guidobaldo del Monte shows this kind of three-dimensional method being applied to the problems artists had traditionally dealt with in two dimensions (or had solved with string). We are at last seeing the three-dimensional 'pyramid of vision', with its vertex in the eye. Moreover, as we shall see in the next chapter, these

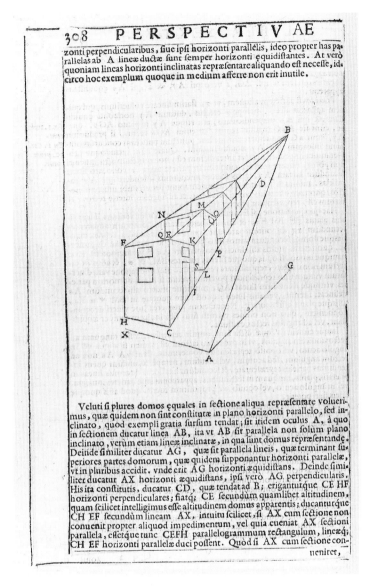

Fig. 7.30 Guidobaldo del Monte, *Six books on perspective* (Pesaro, 1600), p. 308, diagram to illustrate Book 6, 'On stage scenery', showing a block of houses. The eye is at *A*. Compares this with the scenery of the Teatro Olimpico, shown in Figs 6.22 and 6.24 to 6.26.

same authors think in rather the same way about the relation of conic sections to the cone. They do not, however, put things together in a way that shows up the analogy between the two configurations. This limitation of their thought may be due to the tendency we have already noted in the work of Benedetti, namely the concern to provide specific solutions to specific problems, rather than to explore the method of solution for its own sake. Abstract and patrician though the work of Commandino, Benedetti and Guidobaldo may seem, it turns out not to have been quite abstract enough.

# Chapter 8

# BEYOND THE ANCIENTS

Renaissance painters found themselves with very little ancient material that they could copy more or less directly. Sculptors had considerably more, and the quantity was steadily increased as excavations provided collectors with further pieces. Architecture survived in larger quantity, though mainly in a ruined state that left a certain amount of scope for the historical imagination, which of course could be aided by surviving texts. In texts, as well as in archaeological survivals, it was architects and engineers who were best provided with evidence about ancient practice. Only for the products of the minor arts, such as medals, cameos and small sculptures, was there a greater mass of evidence from surviving objects than from written sources. For such objects, direct emulation was possible, and sometimes extremely successful in artistic terms (see Fig. 8.1). Indeed, cameos in particular often pose problems for museum curators because it can be very difficult to tell whether some of them were made in Antiquity or in the Renaissance.

The survival of ancient texts is less random than that of ancient artefacts. Even famous bronze statues were melted down when their condition deteriorated. Useful texts, such as that of Euclid's *Elements*, were copied again and again, and thus survived. In fact, the greatest losses of texts from the ancient world were in two large scale acts of deliberate destruction. The first was the burning of the library at Alexandria by a Christian mob in AD 415 (for good measure they also lynched one of the famous philosophers and mathematicians of the day, Hypatia, the daughter of the mathematician Theon of Alexandria). Some of this library, the largest in the ancient world, had already been burned (accidentally) when Julius Caesar was besieged in the city in the first century BC. The second, and probably more serious, act of destruction was the sack of Constantinople when Venetian persuasion turned the Fourth Crusade against the Byzantine Empire in 1204. Though they did not think of it that way, the learned Venetian humanists of the Renaissance can be seen as making a gesture of atonement on behalf of their ancestors.

Despite these acts of vandalism (which we may note were not actually the responsibility of the Vandals) we still have at least large parts of most of the mathematical works the ancients themselves considered to be among their greatest intellectual achievements. The survival rate for the material monuments known as the Seven Wonders of the ancient world is more like one in seven, namely the pyramids of Egypt. Though texts sometimes seem to have dropped out of sight—and the Renaissance saw the triumphant rediscovery and publication of a number of such works—there is a high degree of continuity in the learned tradition of mathematics.

Renaissance scholars were very much aware of this continuity. For instance, Nicolaus Copernicus (1473–1543), who received a good humanist education at the universities of Krakow, Padua and Ferrara (where he took a degree in law), is now best remembered for his innovation in proposing a Sun-centred model of the Universe. However, in putting forward this system, Copernicus takes great care to point out that he has ancient predecessors, though their writings

**Fig. 8.1** Antico (*c.* 1460–1528), *Mercury*, bronze partly gilded, height 41 cm, *c.* 1500, Victoria and Albert Museum, London. The sculptor, whose real name was Pier Jacopo Alari Bonacolsi, was known as Antico because of the authentically classical appearance of his works.

have not survived. In fact, modern scholarship does not completely endorse his account of the matter, but he certainly did have one ancient predecessor, namely Aristarchus of Samos (*c.* 310–*c.* 230 BC). The case of Commandino's edition of Ptolemy's *Planisphere* shows that the search for ancient antecedents in particular cases could sometimes lead to new mathematical insights. As we shall see, however, the same pressure did not operate to such good effect in regard to Commandino's engagement with work on conic sections. There the connection with ancient work was plain to see, in the famous treatises of Archimedes and Apollonius of Perga, and the standard references to a 'cone of vision' in works on optics somehow did not give any sixteenth-century mathematician the idea of considering the properties of conic sections in terms of perspective.

## Conics in practical texts

This failure (perhaps one should merely call it an omission) to put things together in a way that, from the vantage point of the twentieth century, seems entirely obvious, may be partly explained by the fact that conics belonged almost exclusively in the world of learned mathematics. Treatises on practical mathematics that deal with perspective rarely mention conics—except, of course, the circle. Other conics were considered difficult. Moreover, they were not obviously useful things to know about. This is made clear, though doubtless unintentionally, in one of the rare appearances of conics in a practical treatise, namely Albrecht Dürer's discussion of them in his *Treatise on measurement …* (1525). This work seems to have been widely read. It came out in Latin translation, for an international readership, in 1530, under the emended and rather grander title *Geometry*.

Dürer considers all the conics, but the practical connection seems to be that the parabola gives the shape for making a burning mirror (see Fig. 8.2). He shows the curve as having the property that rays parallel to its axis will be reflected to one particular point. The mathematics of this could be found in several ancient texts, but Dürer gives neither references nor proof. Instead he merely says 'The cause of this has been explained by mathematicians'. Moreover, we know enough about mirror making in the Renaissance to be sure that in practice the shape of burning mirrors was not parabolic but spherical. A segment of a sphere is close enough to the corresponding part of the paraboloid to make an effective burning mirror—and the quantity of radiation from the Sun is sufficient to make high accuracy in the reflecting surface an unnecessary luxury. (It is, of course, a different matter if one wants to make a parabolic mirror that will give one good images of stars, but that was an eighteenth-century problem.) So in practical terms Dürer is providing theory for consumption by the patron rather than for use by the artisan.

He does, however, suggest a neat practical method of drawing conics. As can be seen in the diagrams for the parabola, this involves drawing a vertical section through the cone and then drawing the horizontal section at a series of points suitably spread out along the line representing the cutting plane (that is the plane of the conic). The horizontal sections are superimposed, with a common centre, and on each of them the edge of the part of the cone below the cutting plane is shown as a circular arc, struck with compasses from the common centre. Lengths are then taken over with compasses, effectively giving coordinates for points on the conic. This looks as if it were a good method for drawing conics, but its accuracy in practice is called into question by Dürer's illustration of the ellipse, shown in Fig. 8.3. His ellipse is egg-shaped. On some occasions one may blame the illustrator, but we happen to know that for this book Dürer was his own illustrator, so what we

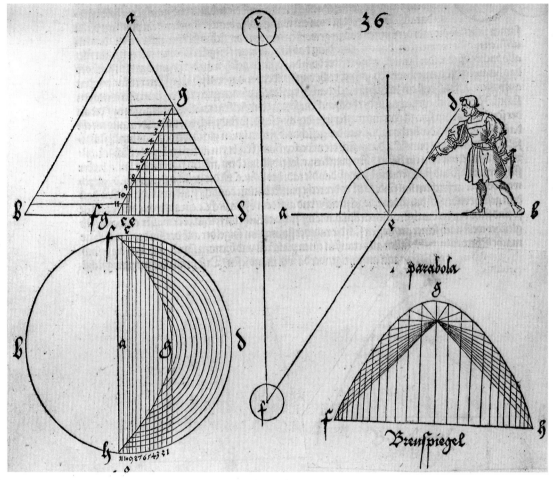

**Fig. 8.2** Albrecht Dürer, *Treatise on measurement …* (Nuremberg, 1525), Fig. 36, the parabola as a burning mirror, and a method of constructing the curve by use of vertical and horizontal sections.

are seeing is presumably in accordance with his intentions. He seems to have believed that the curve obtained by cutting a cone in this way was not symmetrical. This in turn suggests that he did not know, or did not accept as accurate, the method of drawing an ellipse with a thread looped round two pins, since this method clearly will give a symmetrical curve (see Fig. 8.4). Dürer was by no means alone in taking the ellipse to be egg-shaped. The confusion was perpetuated by the habit of calling the curve an 'oval' (*ovalis*), which literally means 'egg-shaped'. This term is even used a few times by Kepler, who certainly knew the curve was symmetrical (see below).

## Commandino and the ancients

It is probably purely as a matter of convenience that Dürer has drawn the cones as different shapes, that is with different vertical angles, for the ellipse and the parabola. However, Archimedes presents each conic as a section of a differently shaped cone, since the plane of section is always to be

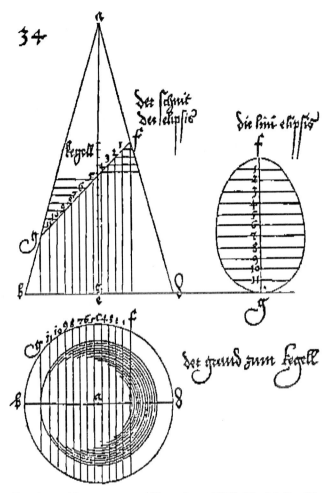

**Fig. 8.3**  Albrecht Dürer, *Treatise on Measurement* … (Nuremberg, 1525), Fig. 34, the ellipse, and how to draw it.

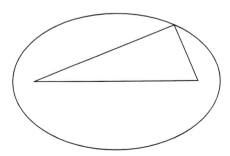

**Fig. 8.4**  Drawing an ellipse by means of a loop of string passed round two pins.

perpendicular to a generator (shown as *ad* in Dürer's figures). Thus the parabola, in which the plane of section has to be parallel to a generator (it is parallel to the line *ab* in Dürer's figure), is presented by Archimedes as the section of a right-angled cone. (The names ellipse, parabola and hyperbola relate to a different series of properties of the curves, and are derived from Apollonius.)

Commandino's Commentary on Ptolemy's *Planisphere* was published in the same year as his edition of the work of Archimedes (1558). This may partly account for the fact that when Commandino looks at various sections through the cone his method of investigation is essentially Archimedean. He looks for conditions on the cone which will ensure that the second section obtained by the projection will have the required shape. In the stereographic projection, which is the one Ptolemy is concerned with, the section will be circular. In any case, Commandino's approach ensures that each cone he examines will contain one circle plus only one other conic section. Like Archimedes and Apollonius he is essentially dealing with each conic separately. Guidobaldo's treatment of conics in his *Six books on perspective* is exactly like Commandino's in its content, though considerably longer and more detailed in its style.

Unlike Archimedes, Apollonius does construct all his conics as sections of one cone, and it is the mathematicians' true cone, extending on both sides of its vertex, so (unlike Dürer and many others) Apollonius does obtain both branches of the hyperbola. However, having obtained the conic sections, Apollonius proceeds to forget about the cone and treat each conic as a plane curve. He shows up the relationship of one conic to another by proving that the curves have analogous properties. For instance, for an ellipse the sum of the distances from a point on the curve to its two foci is constant (hence the method of constructing the curve shown in Fig. 8.4), whereas for a hyperbola it is the difference of the two distances that is constant, and for the circle the distance itself is constant. Apollonius' analogies lead him to a coherent theory of conic sections, but each theorem of the sets of analogous results is proved separately. What he has is a coherent theory but not a truly unified one.

## Kepler on conics

It is a matter of taste how unified one likes one's theories to be. Johannes Kepler's taste was for as much unity as possible. That was one of his reasons for accepting the heliocentric theory of the planetary system. In the heliocentric theory, all the one-year cycles in the paths the planets traced out against the pattern of the fixed stars could be explained in exactly the same way, that is as due to the motion of the Earth round the Sun. For Kepler, metaphysical prejudices came first (he put them into print in 1596), and the long slog of using Tycho Brahe's observations to prove he was correct came later (from 1600 onwards).

Kepler's discussion of conics is in connection with optics, in a treatise with the cumbersome title *Things not in Witelo with which the optical part of astronomy is concerned* (*Ad Vitellionem paralipomena quibus astronomiæ pars optica traditur*, Frankfurt, 1604). (The title is confirmation that Witelo's work on optics, written in thirteenth century, was the standard textbook on the subject in Kepler's time.) Kepler was, in principle, trying to explain the functioning of the *camera obscura*, which was used in making observations of the Sun and Moon (see Chapter 7). His particular concern was to find out how the size of the aperture affected the size of the Solar image. This problem had proved less simple than it may sound, and Kepler was in fact the first person to get the right answer. On his way to it, he gave the first correct description of the functioning of the human eye and made an attempt to give a greater degree of unity to the theory of conics. The conics had entered the story because of their connection with burning mirrors.

Kepler's suggested system of conics is shown in Fig. 8.5. He has put them all together in the same plane, with a common axis, *fr*, and a shared focus at the point with the four letters *a, c, d, e*. ('Confocal' conics in the twentieth-century sense would share both foci.) The system is fairly complicated and depends upon the idea that the properties of conics will change slowly from one type to the next. For instance, as the foci move closer together an ellipse comes closer and closer to being a circle. Also, as the ellipse becomes longer and longer, its foci moving further and further apart, it comes closer and closer to the parabola. So where is the second focus of the parabola? Actually, Kepler seems to have come to his answer to this question from a different direction, namely by considering the conics as two-dimensional versions of burning mirrors.

Kepler's definition of a burning mirror is that light emitted at one focus will be reflected to the other. As he points out, the relevant reflection property for the ellipse (shown in our Fig. 8.6) is proved by Apollonius. For the parabola (see Fig. 8.2) light emitted at the focus shown as *e* in Kepler's system will be reflected parallel to the axis (the line *fr*). In the left half of his diagram Kepler shows a reflected ray, the straight line *gh*, *gi* which goes to what he calls the 'blind' (meaning 'invisible'?) focus of the parabola. That is, the second focus lies in *both* directions along the line *hgi*. It has to be in both directions in order to provide continuity with both the ellipse and the hyperbola. And it clearly lies at what Kepler calls an 'infinite' distance. Unfortunately, the Latin word he uses to describe the distance could also be translated 'indefinite', so we must not put too much weight on the actual name 'point at an infinite/indefinite distance'. Mathematicians and philosophers of Kepler's day usually kept well clear of discussing things that might be indefinitely

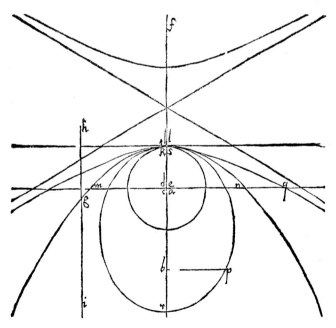

**Fig. 8.5**  Johannes Kepler, *Things not in Witelo ...* (Frankfurt, 1604), p. 94, showing a plane system of conic sections.

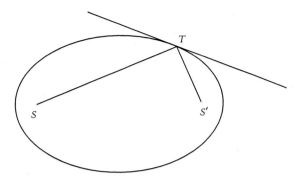

**Fig. 8.6** Ellipse with foci *S* and *S'*. The lines *ST* and *S'T* make equal angles with the tangent to the curve at *T*.

(or infinitely) large, so there was no precise vocabulary available. Kepler accordingly seems to flail about a bit in explaining what he means, but the fact that the second focus of the parabola is said to lie in both directions shows that it is in fact the same as what, following the work of Desargues, became known as a 'point at infinity'.

When he first discusses such points, in his *Rough draft on conics* of 1639 (see below for the hugely long full title of the work), Desargues cites nobody. In fact, he rarely cites predecessors in any of his works. This is unhelpful to historians, but would not necessarily have been considered improper at the time. Desargues' approach to points at infinity is quite different from Kepler's, so perhaps we should leave the matter there. But, as so often, there is an irritating piece of counter-evidence, this time to suggest Desargues could have known Kepler's work. In the course of a massive treatise on music, which includes a few pages written by Desargues, Marin Mersenne (1588–1648), who was a personal friend of Desargues, discusses the reflection of sound, and reproduces Kepler's diagram of the system of conics (see Fig. 8.7). The copy is so close that it seems Mersenne's draughtsman must have had a copy of Kepler's book open in front of him. We have in fact got a case of 'spot the difference': Mersenne's version leaves out the line parallel to the axis going to Kepler's point at infinity. However, Mersenne might well have lent Desargues his copy of

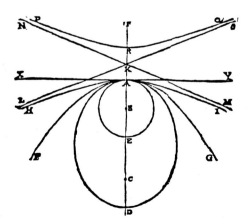

**Fig. 8.7** Marin Mersenne, *Universal harmony* (*Harmonie universelle*, Paris, 1636), Book 1, 'On the nature and properties of sound', p. 62, system of conics (cf. Fig. 8.5).

Kepler's book. The geometry it contains is interesting, and Desargues was interested in geometry. In any case, Kepler's treatise seems to have been well known in learned circles, in France as elsewhere, chiefly on account of its convincing discussion of the working of the eye.

## Sundial problems

Although there are ancient works that discuss conics in connection with burning mirrors, their more usual context is that of sundials. The mathematics, which everyone before the mid-seventeenth century seems to take as too obvious to need explaining, is as follows.

Forget this Copernican (or Aristarchan) nonsense about the movement of the Earth. What we see is the Sun going round the Earth once in a day, following a path that is very nearly an arc of a circle, the tiny departure being due to the fact that day by day the (roughly East) point at which the Sun rises moves along the horizon a little, as does the (roughly West) point at which the Sun sets. Now imagine a line joining the centre of the Solar disc to the tip of a pointer. This line has one fixed point, the tip of the pointer, and one point that moves round a circle, the centre of the Sun. So, as the Sun moves, the line will sweep out part of the surface of a cone. Accordingly, the line this moving line will trace on a plane will be the curve in which that plane cuts the surface of a cone, that is, a conic section. What kind of conic it will be will depend on the relation of the plane (that is the surface of the sundial) to the plane of the apparent motion of the Sun. Thus, the shadow of the tip of the pointer will always move along a conic in the course of the day, but as the plane of the motion of the Sun changes (as it does, slowly, through the year) the conic will change also.

Consequently, the dial plate of the sundial will need to have not one line but a series of lines for the passage of the tip of the shadow of the pointer (which is called the 'gnomon' of the sundial—not to be confused with the other kind of gnomon mentioned in Chapter 7). Each line will correspond to a particular date, and each will have the form of an arc of a conic section. That is, lines of this form will be required if the plate of the sundial is to be flat. The mathematical problems are much simpler if one adopts the usual ancient design for large fixed sundials and draws the lines on the inside of a sphere. However, the Greeks and the Romans did make flat sundials, and the method of drawing their lines is discussed at some length and in some detail by Vitruvius. What he says is not very clear, and he explains nothing of the underlying theory. Probably it would have been inappropriate for him to attempt to do so. The general level of understanding of such things is indicated by the fact that it apparently came as a surprise to the political authorities when a sundial captured in the sack of Syracuse (in Sicily, latitude about 37°), in 212 BC, failed to work properly in Rome (latitude about 42°). Moreover, Seneca (d. AD 65) left on record the acid comment that it is easier to get agreement among philosophers than among clocks.

Despite recent improvements in mechanical clocks (a medieval invention), sundials were still of practical importance in the Renaissance. One used a sundial to reset the mechanical clock when it either stopped or was obviously telling the wrong time. Thus the Renaissance interest in Vitruvius' discussion of sundials was not mere antiquarianism. On the other hand, there is certainly an element of the parade of learning in the fashion for very elaborate sundials which is characteristic of the sixteenth and seventeenth centuries. An example is shown in Fig. 8.8.

**Fig. 8.8** An elaborate sundial in the shape of a polyhedron (a rhombicuboctahedron), brass, partly gilded and silvered, height 13 cm, diameter 6.3 cm, late sixteenth century, probably German, British Museum, London.

Vitruvius' drawing instructions are not the only writing on sundials to survive from ancient times. There is also a short mathematical treatise by the astronomer Claudius Ptolemy, called *On the analemma*. Commandino duly made an edition and translation of it which were published in Rome in 1562. From the point of view of a mathematician, this is a vastly more interesting text than Vitruvius'. Commandino used it as a starting point for a very detailed exposition of the theory of sundials which was published together with his edition of Ptolemy's work. Commandino is obviously setting out to explain the mathematics behind the different types of dial plate described by Vitruvius. Many of his excellent diagrams, and a competent adaptation of his explanation, appear in the commentary notes to the second edition of Daniele Barbaro's Italian version of Vitruvius' *On architecture*, and in his edition of the original Latin text (both published in Venice in 1566). Sundials were thus firmly ensconced as a part of mathematical practice that was of interest to the learned.

## Benedetti on conics

Many learned treatises on sundials were published in the sixteenth and seventeenth centuries. One of the clearest, and among the most widely read, was the treatise by Giovanni Battista Benedetti, *A book on the use of gnomons and solar shadows* (*De gnomonum umbrarumque solarium usu liber*, Turin, 1574). Like Benedetti's treatise on perspective, the sundial book is well supplied with inelegant but serviceable illustrations. The problems are, moreover, treated in a manner that would allow one to carry out the corresponding procedures in practice.

One part of the book, however, does not look very practical. This is the Appendix, which describes a new instrument for drawing the arcs of conics that will be required. Benedetti's picture of the instrument is shown in Fig. 8.9. At the time, the usual way of drawing conics was to use an

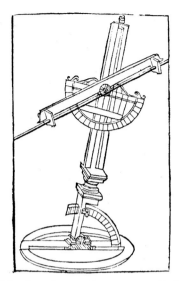

**Fig. 8.9** Giovanni Battista Benedetti, *A book on the use of gnomons ...*, Turin, 1574, Appendix, f. 117 *verso*, instrument for drawing conic sections.

apparatus called trammels whose design was based on plane properties of the curves. This apparatus was known to medieval Islamic mathematicians and is illustrated in Dürer's *Treatise on measurement*.... It does not usually work very well. The design gives various sliding pieces ongoing possibilities of getting jammed. So a new apparatus for drawing conics was, in principle, a welcome invention. In fact, Benedetti's instrument does not look likely to be better than trammels, and there is no evidence the instrument was ever made. But its mathematics is interesting.

As can be seen in Fig. 8.9, Benedetti's instrument works from above the plane of the section (which is horizontal), and effectively sets up the axis and vertex of the cone through which the section is being taken. To choose which conic we draw, we need to know how the plane in which such a conic will appear is related to the axis and vertex of the cone. Some of the theorems Benedetti needs to make his choices are in Apollonius, to whom Benedetti duly gives references. However, Benedetti is working in three dimensions, whereas Apollonius prefers to work in two. As a result Benedetti comes up with several theorems that are not in Apollonius' *Conics*—not even in the later books of it which, at the time Benedetti wrote, had not yet been recovered. The diagrams corresponding to two of these new theorems are shown in Figs 8.11 and 8.12. Benedetti's text makes it clear that the bases of his cones are circular and that the sections shown in the cones are ellipses. Unfortunately, the draughtsman has lowered the intellectual tone by making his drawings of the ellipses lenticular. This is a fairly common way of showing circles in perspective in book illustrations in the sixteenth century, so it is not clear whether one should regard it as a howler or a drawing convention.

Benedetti's fourth proposition is that if a cone is cut by two planes parallel to one another, then the two conic sections that result will be similar (see Fig. 8.10). That is, in the case shown in his diagram, the curves will be ellipses whose axes are in the same ratio. For this result, Benedetti gives a reference to Apollonius. He next considers what will happen if we change the angle at *a*, that is the 'vertical angle' of the cone (see Fig. 8.11). This theorem is one of several indications that Benedetti has read not only Apollonius' but also Archimedes' works on conics. He proves that if we have two cones with different vertical angles, then a plane of section passing through both

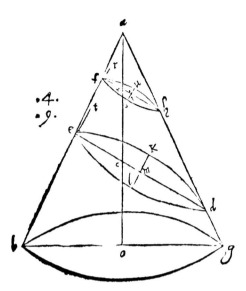

**Fig. 8.10** Giovanni Battista Benedetti, *A book on the use of gnomons …*, Turin, 1574, Appendix, f. 118 *recto*, cone with two parallel sections, giving similar ellipses.

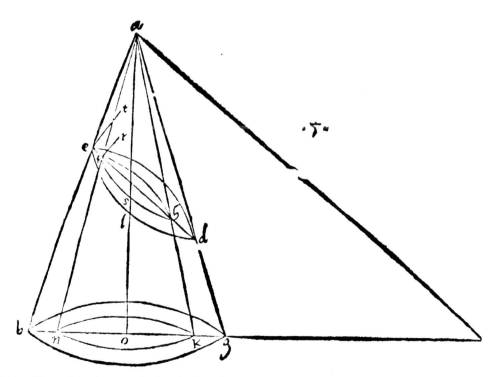

**Fig. 8.11** Giovanni Battista Benedetti, *A book on the use of gnomons …*, Turin, 1574, Appendix, f. 119 *recto*, two cones with a common vertex, at *a*, showing the two different ellipses given by the same plane of section.

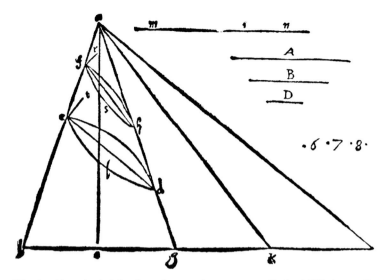

**Fig. 8.12**  Giovanni Battista Benedetti, *A book on the use of gnomons …*, Turin, 1574, Appendix, f. 120 *recto*, a cone cut by planes that are not parallel to one another, giving two different ellipses.

cones will give us two different conics. In the case shown, the curves are ellipses whose axes are not in the same ratio to one another. In his next proposition, Benedetti shows that the ratio between the axes of the ellipse also depends on the angle between the plane of section and the axis of the cone (see Fig. 8.12).

The theorems illustrated by the diagrams in our Figs 8.11 and 8.12 are not in Apollonius. Nor are the following series of problems, by which Benedetti establishes how to determine exactly what angles must be set up on the two circular scales of his instrument in order to draw a conic whose axes are in some given ratio or of some given length. Benedetti's new results clearly belong to the instrument, and we have no indication that they were initially conceived separately from it.

The proofs of the new results are all quite simple, and the results themselves clearly arise from the fact that, unlike Apollonius, Benedetti has not discarded the cone. His proofs in fact depend upon considering the properties of triangles in the plane defined by the axis of the cone and an axis of the conic, a plane that is perpendicular to the base of the cone. As we shall see, it is triangles just like these that play a crucial part in Desargues' treatment of conics in 1639.

## Girard Desargues

In his work on conics, as in his work on perspective, Benedetti is conceiving problems in three-dimensional terms, exactly as Desargues was to do about fifty years later. The approach is not like that of ancient geometers, and it brought Benedetti some new results, though it did not lead him to a general theory like that put forward by Desargues. The great difference between the two men is that Benedetti, despite being a highly competent and original mathematician, appears to be unaware of the power of his new method, whereas Desargues does recognise the power of the method and goes on to give a projective treatment of conics. It is, in fact, not only in his work on perspective and conics that Benedetti behaves in this way. Rather similar judgements could be made of his work in other areas, such as music theory and the study of motion. In all these areas, what seems to be happening is that Benedetti has set out to solve some specific problems suggested by practical concerns, and has solved them, but has not then looked into any further developments that might be suggested by the methods he has used in obtaining his solutions. This may seem puzzling psychologically, since the methods are highly original, as Benedetti certainly knew, but in historical terms all one can really say is that Benedetti epitomises the sixteenth century's failure to be the seventeenth. The original ideas arising out of practical mathematics were not seen as having importance outside that tradition.

Part of the reason for the very different attitudes displayed by Desargues may be that he, even more than Benedetti, was something of an outsider. Like his near contemporary René Descartes (1596–1650), Desargues had no need to earn his living, and wrote essentially only to please himself and his friends. In fact, when it came to patronage, Desargues seems to have been the patron rather than the client. This information has come to us rather recently, from work in the legal archives in Desargues' native city of Lyon, and in Paris. The first results of these investigations were made public in 1991.

It had always been known that Desargues came from Lyon, since his writings identify the author as L.S.G.D.L., for 'Le Sieur Girard Desargues Lyonnais'. The legal archives provide much additional information. In particular, they record Girard's protracted attempts to recover a huge sum of money owed to his father by a former Tax Farmer. In France at this time, the King did not collect taxes directly through his officials. Instead, a rich citizen paid the King an agreed sum of money, in principle equal to the amount of tax that was due, and this person, called the Tax Farmer, then set about collecting the taxes to recoup the sum concerned. In order to pay the King, the Tax Farmer usually had to borrow money, in large quantities. It is clear that since Girard Desargues' father was involved in this kind of high finance he must have been one of the richest private citizens in Lyon, which is to say he was among the richest men in France. Moreover, even after the Tax Farmer had, apparently, reneged or stalled on the repayment of his debt, the Desargues family was still very wealthy. We know that it owned a manor house and land in a large village called Vourles, about ten kilometres from Lyon, a small château nearby with one of the best vineyards in the region, and one of the largest houses in the city of Lyon, as well as various other smaller properties in the city and elsewhere. This wealth had apparently been built up over several generations. Moreover, members of Girard's mother's family had been prominent in learned legal circles for about five generations, in Lyon and in Paris.

These discoveries in the archives help to make sense of our previously rather fragmentary information about Girard Desargues himself. For instance his writings make it clear that he knew a great deal about the practical activities associated with engineering, building works and architectural design. There is, however, no hard evidence that he ever practised as an engineer or as an architect—for example, we have no contracts, and no records of payments. Now we know Desargues was very rich this lack of evidence is easily explained: he did not earn his living in this way. His interest in such matters was rather that of a patron than a practitioner. Due allowance being made for differences of time and place, Desargues now looks very like Daniele Barbaro—the huge difference being that Desargues was incomparably more talented as a mathematician.

Another aspect of Desargues' work that is now easily explained is his apparently very thorough acquaintance with the standard works of the learned mathematical tradition, particularly the works of Pappus of Alexandria and Apollonius of Perga. It had seemed a bit awkward to account for such knowledge on the part of someone who was apparently of lowly social standing, with no known patron, and who wrote not in Latin but in French. It now seems obvious that Desargues received a standard humanist education and could read Latin (and possibly also Greek). There is no question that he could afford to buy what books he pleased, though the inventory taken in one of his houses at his death (1661) unfortunately refers only to furniture and works of art, not to books. The references to works of art do not give any detail, but it is clear that some paintings, at least, were there for their own sake rather than merely for devotional purposes.

The family's wealth also explains why contemporaries quite often convert the name into more aristocratic forms, such as De Sargues, Des Argues, Du Sargues. Similar operations are performed on the name of Descartes. These presumably account for the fact that the French adjectives corresponding to the two names are formed similarly: 'cartésien' and 'arguésien', though, for reasons we shall discuss later, the latter adjective is generally more useful for games of Scrabble than for writing histories of mathematics.

## Desargues on perspective

Desargues' intellectual life has to be reconstructed from his published writings, and from a few references to him, or them, in the publications and private letters of others. His most important work, the treatise on projective geometry, appeared in 1639, when Desargues was 48. It has a long and slightly odd title: *Rough draft for an essay on the results of taking plane sections of a cone* (*Brouillon proiect d'une atteinte aux evenemens des rencontres du cone avec un plan*). The only work Desargues had published before this, apart from the few pages on music in Mersenne's *Universal harmony* (1636) which we have already mentioned, was a very short book on perspective, with the long and slightly odd title: *Example of one of S.G.D.L.'s general methods concerning drawing in perspective without using any third point, a distance point or any other kind, which lies outside the picture field* (*Exemple de l'une des manieres universelles du S.G.D.L. touchant la pratique de la perspective sans emploier aucun tiers point, de distance ny d'autre nature, qui soit hors du champ de l'ouvrage*). This treatise is only twelve pages long and consists of a single worked example, illustrated by two high quality engravings, shown in our Figs 8.13 and 8.14. These plates are almost certainly by the famous engraver Abraham Bosse (1602–1676), who had a long association with Desargues. As we shall see, Bosse wrote expanded versions of several of Desargues' works. In particular, we shall have more to say below about Bosse's expanded version of the mini-treatise on perspective, which was published in 1648. Bosse in fact appended to this work a reprint of Desargues' original twelve-page book of 1636—in vindication of the originality of the method Desargues had proposed, in connection with which there had been a long-running priority dispute. The right of the matter seems to have been on Desargues' side, though that is not in itself a matter of great importance (see Chapter 9).

What was new about Desargues' method was that, as the title of his book proclaims, it enabled one to work entirely inside the picture field. Alternatively one could use a separate piece of paper, as shown in the scrolls inset at the top of Desargues' first plate. The principle is that one constructs scales which give the progressive diminution in width and height when lengths are seen at greater distances from the picture plane, for some given position of the eye. The method of constructing the scales is, naturally enough, highly reminiscent of all the other methods of perspective construction that have been described in our earlier chapters. One sees again the patterns of pencils of lines radiating out from the foot of the perpendicular from the eye to the picture plane, and from another point whose position is determined by the distance of the eye from the picture. As a

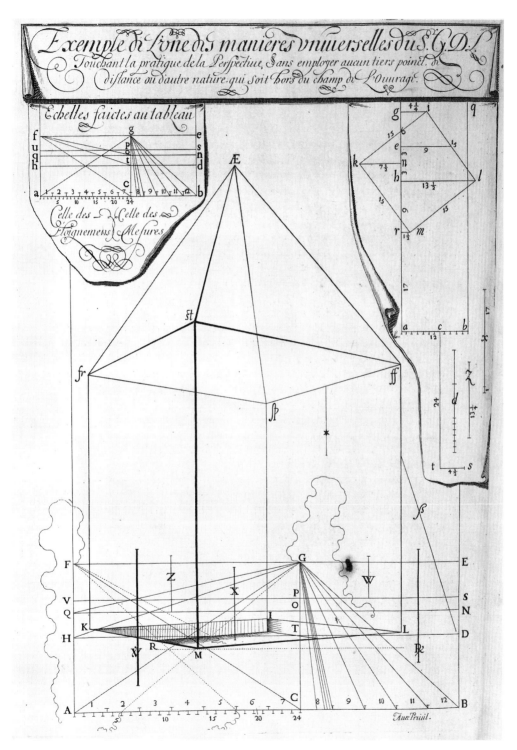

**Fig. 8.13** Girard Desargues, *Example … concerning drawing in perspective …*, Paris, 1636, engraved plate showing the pavilion whose image is to be drawn, and the preliminary construction for the scales of height and width.

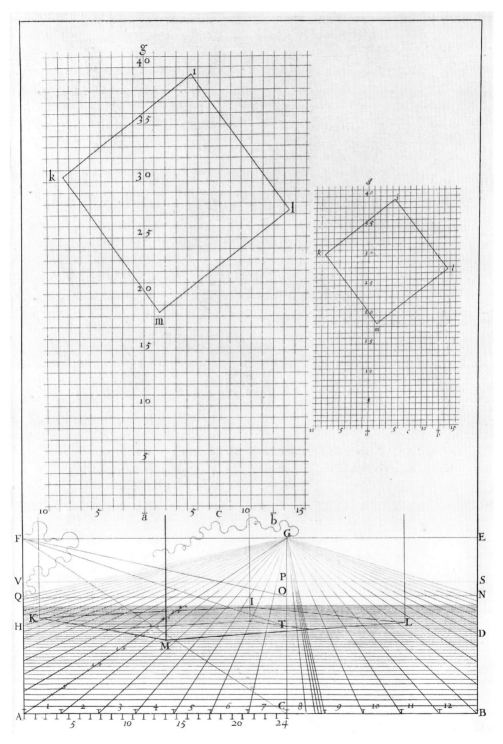

**Fig. 8.14** Girard Desargues, *Example … concerning drawing in perspective …*, Paris, 1636, engraved plate showing the ground plan of the pavilion whose image is to be drawn and the use of the scale for heights.

further reminiscence of practical treatises we see the curly ends given to lines to show that they indicate not actual lines but the positions of threads used to construct them.

There was no standard vocabulary for writing about perspective, so Desargues lists the terms he proposes to use. These appear mainly to be adapted from Guidobaldo del Monte. Desargues also mentions where his own words differ from those used by others, which makes it clear that he has read a great deal of the standard literature on perspective. The fact that he gives no proof to show how his new method works suggests he expects his readers to have done the same (or not to care).

It is not really clear to what kind of readers Desargues is addressing himself. The work follows the pattern of practical treatises by showing a worked example, with actual numerical values given for the various lengths. However, since there is only one example, and it has been chosen to be as general as possible, it is so complicated that what we have is almost a parody of a practical treatise. It is difficult to believe Desargues really expected artists in general to have a sufficiently good grasp of mathematics to see how the technique could be applied in other cases. It seems more likely that he was essentially writing for mathematicians—that is, for the Guidobaldo-reading classes, to whom a number of accounts of perspective were addressed in the early years of the seventeenth century. All the same, Desargues' new method was useful in practice (which is why there was a priority dispute). It was in fact the first of a number of 'abbreviated' methods which allowed work to be carried out only within the picture field.

Though one cannot be sure of the intended readership of the main body of Desargues' text, there is no ambiguity attached to its final page, which is explicitly addressed to theoreticians (*contemplatifs*). One might perhaps have expected a proof of the correctness of the preceding construction. Desargues does not give one. Instead, without warning, he picks up Guidobaldo's results about sets of parallel lines, and makes them look much more interesting. He gives a series of results concerning the perspective images of sets of lines that all converge to a point and sets of lines that are parallel to one another. As in the work of Guidobaldo, the cases considered involve different relationships between these 'subject' lines and a line through the eye (the line of sight). Desargues gives strictly analogous accounts of sets of lines that meet and sets of lines that are parallel. He does not develop the analogy, but clearly intends to put it on display. Perhaps at this time Desargues himself had not thought the matter through any further than this. Three years later, in his work on conics, he states that a set of parallel lines is a set of concurrent lines whose meeting point lies at an infinite distance, and then goes on to give a rigorous discussion of the mathematical properties of the 'point at infinity' at which such parallel lines meet. This later discussion makes no mention of perspective, but the nature and context of the earlier one suggests very strongly that there was a connection. The impression is strengthened by the abrupt and somewhat cryptic reference to conics—the subject of the 1639 treatise—in the final paragraph of the work on perspective:

Given to portray a flat section of a cone, draw two lines [in it] whose images will become the axes of the figure which will represent it.

Desargues does have the grace to preface this enigmatic throwaway with the comment that 'the proposition which follows cannot be explained as briefly as the preceding ones'. However, that is absolutely all he volunteers by way of explanation. This seems a bit unfair even in the twentieth century and must have seemed considerably more so in the seventeenth. In 1636 there were no precedents (that I know of) for considering the perspective image of a general conic.

# Desargues' projective geometry

The hints thrown out in 1636 are open to a number of interpretations, but they certainly indicate that at least some of the ideas we find in the treatise of 1639 had been in Desargues' mind three years earlier, in the context of perspective. This is some comfort to the would-be historian, because the 1639 work presents itself, in the manner of the *Elements*, as being context-less absolute truth. Desargues explains his definitions as little as Euclid does. Ideas are simply put before us. Some of them are merely unconventional, for instance:

In this work, every line is, if necessary, taken to be produced to infinity in both directions.

And again

In this work every Plane is similarly taken to extend to infinity in all directions.

These are now the standard formulations, but they were not so in 1639, and some readers may have winced at the apparently casual use of infinity. One of the important things about Desargues' work is that such readers were misguided. Desargues is the first mathematician to get the idea of infinity properly under control. He uses the concept in a completely precise mathematical way. This does not appear immediately, because having introduced infinity Desargues next turns to definitions of very simple mathematical things, whose significance is not at first apparent. For instance:

When through various points of a straight line there pass other straight lines, in any manner, the line on which these points lie is called a *trunk* (*tronc*).

And the points themselves also receive a name:

The points on the trunk through which other straight lines pass in this manner are called *knots* (*nœuds*).

Following definitions also give names that are parts of a tree. For instance, the lines passing through the knots are called 'branches' (*rameaux*); see Fig. 8.15. It is in fact the 'tree' itself that turns out to be the crucial component in Desargues' reasoning, as we shall see in due course.

Before we come to the parts of the tree, there are a number of definitions that give names that belong to the world of engineers. Desargues' very first definitions refer to sets of lines and their points of concurrence:

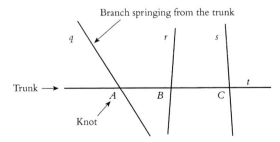

**Fig. 8.15** Knots on a trunk and branches passing through them. The lines *q*, *r*, *s* pass through the points *A*, *B*, *C* on the line *t*. Accordingly, *A*, *B*, *C* are knots, *t* is a trunk, and *q*, *r*, *s* are branches springing from the trunk. Any segment of the trunk lying between two knots is called a branch folded to the trunk.

To convey that several straight lines are either parallel to one another or are all directed towards the same point we say that these straight lines belong to the same ordinance (*ordonnance*), which will indicate that in the one case as well as in the other it is as if they all converged to the same place.

The place to which several lines are thus taken to converge, in the one case as in the other, we call the *butt* (*but*) of the ordinance of the lines.

What Desargues calls an ordinance of lines is now called a 'pencil' and what he calls the butt (meaning its target) is now called the 'vertex' of the pencil. Desargues' terms are more colourful than their modern counterparts and clearly come from the vocabulary of the artilleryman. When the ordinance/pencil is not of lines but of planes, that is when all the planes pass through a single line, the butt/vertex becomes an axle (*essieu*). We are still in the world of the engineer. It thus seems possible that the word *arbre* might also be of related practical origin. *Arbre* is usually translated as 'tree', but the word can equally mean 'arbor' or 'axle'. (In English there is also the term 'axle-tree'.) Like the central axle in a machine, Desargues' *arbre* is the member to which others are related. That is, it is their relation to the *arbre* that chiefly defines their significance in the overall arrangement. The standard metaphorical usage by which engineers called an axle a tree may have suggested to Desargues that the same imagery could provide names for the further, subordinate, elements in his geometrical scheme.

Desargues' vegetable vocabulary, which also shows the vegetable tendency to grow to unexpected size, has attracted much comment. Indeed the words seem to have distracted attention from the substance of what he has to say. In fact, Desargues uses his terms completely consistently, in the normal mathematical way, so that in context they are simply technical terms. Any objection to them can accordingly be overruled by an appeal to the standard definition of a definition:

'But "Glory" doesn't mean "a nice knock-down argument",' Alice objected.

'When *I* use a word,' Humpty Dumpty said in a rather scornful tone, 'it means what I choose it to mean,—neither more nor less.'

*Through the Looking Glass* ... was published in 1872, but Lewis Carroll's formulation applied in the seventeenth century as it did in his own time—and does in ours. All the same, it is not given to all to be as strong-minded as Humpty Dumpty all the time. It is really not clear whether the original readers of Desargues' work found his vocabulary made it less easy to understand. René Descartes did mention the new vocabulary in his carefully phrased suggestion that the work need to be completely rewritten (see below), but most of the recorded seventeenth-century comments differ from those of the twentieth century in concentrating upon the substance of the work.

One awkwardness certainly obtrudes itself upon any reader. Desargues' naming of points in his diagrams is generally untidy, to say the least of it. For instance, among the first results he proves we find Menelaus' Theorem. In the twentieth century this theorem is usually presented as referring to a transversal to a triangle (see Fig. 8.16). Desargues presents it as concerning a transversal to three lines—which effectively comes to the same thing (the three lines being the sides of the triangle). Using the notation of the diagram in Fig. 8.16, in which the triangle is *ABC* and the transversal *PQR*, Menelaus' Theorem states that the relation between the lengths cut off on the sides by the transversal is

$$\frac{QC}{QA} = \frac{PC \cdot RB}{PB \cdot RA}.$$

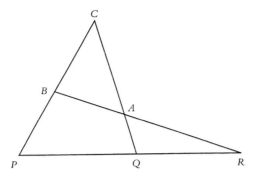

**Fig. 8.16**  Modern diagram for Menelaus' Theorem, with the triangle *ABC* and transversal *PQR*. The theorem establishes a relationship between the lengths cut off by the transversal on the three sides of the triangle.

Desargues chooses to consider lines *hKH*, *h4D*, *K4G* (that is, a triangle *4Kh*) and a transversal *HDG*, so his diagram would be as shown in Fig. 8.17. This diagram is a modern reconstruction, since no diagrams are preserved in the single surviving printed copy of Desargues' treatise. With this lettering, Menelaus' Theorem takes the form

$$\frac{Dh}{D4} = \frac{Hh \cdot GK}{HK \cdot G4}.$$

Confusing lettering may, of course, be a relic of the author's struggle to understand the mathematics concerned, and it could presumably find its way into print if the author was afraid that an attempt to tidy up would introduce mistakes. Desargues' *Rough draft on conics* (as we shall call it, for the sake of brevity) was in fact not a formal and finished publication. There seem to have been only 50 copies printed, destined not for sale to the public but for distribution to Marin Mersenne's circle of mathematical friends and correspondents. Many corrections, some quite important, are made in a 'Notice' added at the end. This style is not found in works that were published in the usual way. What we have seems to be a seventeenth-century equivalent of a typed preprint. So perhaps Desargues' unhelpful lettering/numbering habits are merely testimony to the essentially provisional nature of his text. He did, after all, call it a 'rough draft'.

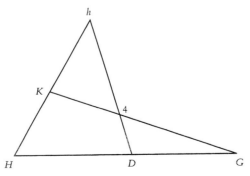

**Fig. 8.17**  Diagram for Desargues' presentation of Menelaus' Theorem. We have lines *hKH*, *h4D*, *K4G* and a transversal *HDG* (compare with Fig. 8.16).

Such an excuse is, however, rather feeble in the case of Menelaus' Theorem. It was a well known result. Desargues correctly points out that it is proved by Ptolemy (in *Almagest*, Book 1, Section 13, though Desargues does not say so). In fact, Desargues' proof is essentially the same as that given by Ptolemy. It has presumably been included in the *Rough draft on conics* because Desargues will have occasion to make repeated use of Menelaus' Theorem. As things turn out, one has plenty of opportunity of learning to recognise the result, however transmogrified by bizarre lettering.

Before coming to Menelaus' Theorem, Desargues investigates some properties of the trunk, that is of what we should now call 'ranges' of points on a line. The most important property he establishes is that of the 'tree'. To state this in Desargues' own terms would involve several more definitions. In modern terms, the three pairs of collinear points $B, H$; $C, G$; $D, F$ form a tree if there is a point $A$ such that

$$AB \cdot AH = AC \cdot AG = AD \cdot AF.$$

The point $A$ is then called the 'stump' (*souche*) of the tree. There are two cases to be distinguished: either $A$ separates the points of each pair, or it does not. That is, in Desargues' terms, the stump is engaged between the knots (as in Fig. 8.18) or it is disengaged (as in Fig. 8.19). The trees that have been shown in our figures are the ones that arise from following Desargues' argument through in detail. Lettering does not come out in alphabetical order. Nor does Desargues write out equations as we have done. Everything is expressed in ordinary sentences (sometimes rather long and complicated ones), and what we have shown as the product of the lengths $AB$, $AH$ is described as the rectangle contained by the line segments ('limbs', *branches*) $AB$ and $AH$. This is the standard Euclidean style, with which all Desargues' original readers would have been familiar.

Having defined a tree, Desargues proceeds to show that the six points, or rather the three pairs of points, may also be defined as having another relationship to one another, this time one that does not involve the stump. If the pairs are $B,H$; $C,G$; $D,F$ as before, then we have

$$\frac{BC \cdot BG}{HC \cdot HG} = \frac{BD \cdot BF}{HD \cdot HF}. \tag{1}$$

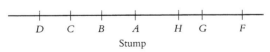

Fig. 8.18 Tree with stump engaged. The pairs of points which form the tree are $B, H$; $C, G$; $D, F$ and the stump of the tree is $A$.

Fig. 8.19 Tree with stump disengaged. The pairs of points which form the tree are $B, H$; $C, G$; $D, F$ and the stump of the tree is $A$.

What this means is that the product of the ratios in which the points $C$, $G$ divide the segment $BH$ is equal to the product of the ratios in which the points $D$, $F$ divide the segment $BH$. It can be shown that a relationship of this form always holds, irrespective of the order in which the three pairs of points are taken. The six points are said to form an 'involution' (*involution*). This is the only one of Desargues' new words that has become part of the technical vocabulary of this kind of mathematics. The involution is not a very simple concept and may not look as if it were interesting mathematically, but it does turn out to be so. This part of Desargues' mathematics is essentially a necessary preparation for the important results that are to follow. It is not of the thrills and spills variety.

Following the standard mathematical ideal of holding out for the maximum degree of generality, Desargues prefers to consider involutions rather than the simpler relation that emerges when the six points are collapsed to four. But he does also consider this case, though he gives no name to the configuration concerned. As can be seen from equation (1), the reduction in the number of points has to be handled a little carefully. If we were to make the points of the pair $C$, $G$ coincide with $D$, $F$, that is identify $C$ with $D$, and $G$ with $F$, then we should simply get the same expression on both sides of the equation—which is then reassuringly true but hardly interesting. What Desargues does is to make each of the pairs $C$, $G$ and $D$, $F$ collapse to a single point. The points $C$ and $G$ are now both $G$, and $D$ and $F$ are both $F$, so equation (1) gives us

$$\frac{BG \cdot BG}{HG \cdot HG} = \frac{BF \cdot BF}{HF \cdot HF}.$$

This is equivalent to

or

$$\frac{BG}{HG} = \frac{BF}{HF}, \qquad (2)$$

$$\frac{BG \cdot HF}{HG \cdot BF} = 1. \qquad (3)$$

As can easily be seen from (2), what this means is that $B$ and $H$ divide $FG$ in the same ratio (internally and externally). This relation is much easier to visualise than that of the involution of six points. The expression on the left-hand side of equation (3), which is the ratio of the ratios in which the points $B$ and $H$ divide $FG$, is now known as the 'cross ratio' of the four points of the range. Desargues does not use this notion (which means of course that he does not give it a name), preferring, as already mentioned, to work with the more general notion of an involution. When he is dealing with cases involving only four points, they are referred to as four points in involution. (In twentieth-century terminology such points are called a 'harmonic range'.)

The work on involutions is followed by the proof of Menelaus' Theorem which we have already mentioned. After this, Desargues reverts to involutions. He first states, and then proves, a result which, once one has disentangled what it means, shows why the involution is important. This result is that if three pairs of points on a line are in involution, then so are their images under projection from a point into another line. Desargues says

When on a straight line $GH$ there are three different pairs of knots $BH$, $DF$, $CG$, which are in involution, and through them there pass three pairs of branches springing from the trunk $FK$, $DK$; $BK$, $HK$; $CK$, $GK$, all belonging to the same ordinance, whose butt is *K*, these three pairs of branches all belonging to one ordinance are, taken together, called a *bough of a tree (ramée d'un arbre)*, and on any other straight line *cb* suitably drawn in their plane, each of them gives one of three pairs of knots of an involution *gh, df, cg*.

Desargues also gives a reference to a diagram. One cannot be sure where the butt *K* was placed in relation to the two lines *GH, cb*. Our diagrams, in Fig. 8.20, show two possible configurations, the first of which has been designed to show up the analogy with the set-up discussed in perspective treatises. We may, if we please, imagine an eye at *K*, looking at the three pairs of points *B, H: C, G; D, F* that are in involution, and seeing their representations in a picture plane passing through *cb* as another set of points in involution.

The pictorial interest of this is, of course, minimal. However, the diagram is rather like a generalised version of the one for the fundamental proposition in the theory of perspective that I called 'Piero's Theorem' (see Fig. 5.3). In Piero's Theorem the second line is parallel to the first, and Piero has shown that the pattern of division (that is, the series of proportions) along the second line is the same as that along the first. Desargues' result involves a way of defining the pattern of divi-

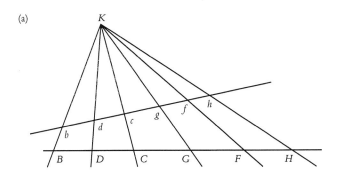

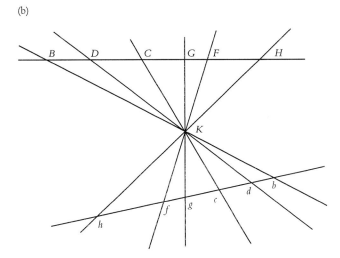

**Fig. 8.20 (a, b)**  Diagrams for Desargues' proof of the theorem that if we have six points in involution, *BDCGFH*, then their images under projection from a point *K* into another line, *bdcgfh*, will also be six points in involution. Either diagram can be used for the proof. Version (a) is like the set-up found in perspective treatises.

sion which allows it to be repeated even if the second line is not parallel to the first one. Mathematically speaking, one could see Desargues as having produced a generalised version of Piero's Theorem. Of course, that is clearly not what happened in historical terms, but the mathematical fact shows up how Desargues is generalising perspective into becoming a technique of use to mathematicians. It also points up the fact that Desargues is not looking at what is changed by perspective, as artists were, but is instead looking for what is not changed. Perspective is not being seen as a procedure that 'degrades' but merely as one that transforms, leaving certain relationships the same. The relations that are left the same are what we now call 'projective properties' of the figure concerned. Thus Desargues' theorem that the six points in involution are projected into another six points in involution could be expressed in modern terms by saying that being six points in involution is a projective property.

Desargues proves his theorem about the involution by repeated use of Menelaus' Theorem. The proof takes most of a page—the pages of Desargues' treatise are folio, with 41 lines of type on each—and involves references to some additional diagrams. The proof is thus a decidedly weighty affair. Desargues is, however, of the opinion that the result is worth the effort, and a modern mathematician cannot but agree—knowing, as Desargues also does, what is to come next. Desargues uses a rather curious metaphor to express this:

The material is rich in similar means of deducing that four points or three pairs of knots on a line are in involution, but the above is sufficient to start off an open-cast mine together with what follows.

Perhaps Desargues' family also had connections with mining enterprises?

What follows, without any warning such as a subheading, is Desargues' treatment of conics. Once again, Desargues strives to be as general as possible, this time by taking into account the possibilities offered by considering points at infinity. For instance, he begins with the idea of a line moving so that one of its points stays fixed and another moves round a circle. He then distinguishes the cases in which the fixed point does or does not lie in the plane of the circle. In considering the more interesting case (the second), he calls the figure that results from it not, as one might have expected, a cone, but a 'roll' (*rouleau*—which means something rolled up, not a bread bun). This is to take account of the possibility that the fixed point may be at an infinite distance, making the roll a cylinder, as well as at a finite distance, making the roll a cone. Desargues emphasises that the true cone lies on both sides of the fixed point, that is on both sides of its vertex. The word 'roll', and Desargues' alternative term 'cornet' for the cone, may seem to originate in the idea of shapes made from a sheet of paper, but we are in fact dealing with abstract geometrical entities. Introducing the general notion of the 'roll' allows Desargues to go on to present sections of a cone and of a cylinder as equivalent.

This part of Desargues' treatise is extremely elegant. It also makes considerable demands upon a reader's powers of visualisation—all in three dimensions. We then return to the involution. After another series of definitions (concerning what would now be called poles and polars) we come to a theorem that is now known as Desargues' Involution Theorem. This is in fact one of two theorems that still carry Desargues' name. The other one, called Desargues' Perspective Theorem, is beautiful and extremely simple. It has already been mentioned in Chapter 7 and will be discussed in our next chapter. The Involution Theorem is not very simple, but it is very important.

To state the Involution Theorem in Desargues' terms, we shall have to go back to some of the definitions given near the beginning of the *Rough draft on conics*. The definitions concerned deal with what is now called the 'complete quadrilateral', that is four points and the three pairs of lines that can be drawn to join them two by two. Desargues calls each of the four points a 'marker post' (*borne*). His definitions are

When in a Plane four points do not all lie in the same straight line we say that each of these points is a *marker post* with respect to the others.

   Each straight line which passes through any two of the four marker posts is called a *marker line* (*bornale droicte*) in relation to these points.

   The two straight lines of which one passes through two of the marker posts and the other through the other two of the four, are taken as forming a pair and are called a *pair of marker lines*.

These terms seem to have been borrowed from surveying—and would have been familiar not only to the surveyor but also to the landowner who employed him. The theorem in which Desargues uses them is, however, thoroughly abstract. It states

When in a plane we have four points *B, C, D, E*, as marker posts paired three times among themselves, through which there pass three pairs of marker lines *BCN, EDN, BEF, BDR, ECR*, each of these three pairs of marker lines and the curved edge of any section of a roll [i.e. a conic] which passes through the four points *B, C, D, E* gives on any other straight line in their plane, such as the trunk *I, G, K*, one of the pairs of knots of an involution *IK, PQ, GH* and *LM*, …

   The least that can be said of this is that Desargues is, as often, trying to do too much at once. He has ended up with four pairs of knots, not the three pairs that are required for an involution. The corresponding diagram would presumably look something like Fig. 8.21.

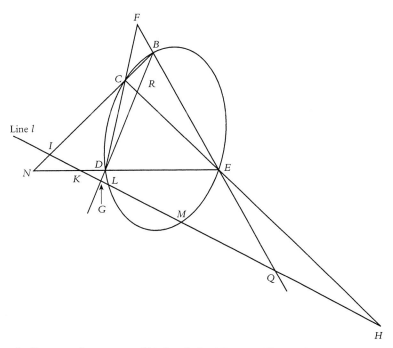

**Fig. 8.21** Diagram for Desargues' statement of his Involution Theorem. The vertices of the complete quadrilateral (marker posts) are *BCDE*. They lie on a conic. Involutions are obtained on the line *LM*.

The proof which follows shows that he is in fact considering two different examples of involutions. The first is the involution produced on a line that cuts the three pairs of sides of the complete quadrilateral. The diagram for this is shown in Fig. 8.22. The three pairs of points in involution are the three pairs of points of intersection of the line we have called *l* with the pair of sides *BC, ED* (that is the points *I, K*), with the pair of sides *BD, CE* (that is *G, H*) and with the sides *CD, BE* (that is *P, Q*). The second example of an involution is shown in Fig. 8.23. Here we have an involution on the line we have called *l* formed by the pairs of points in which it meets any two pairs of sides of the quadrilateral (we have shown the points of intersection with *BC, ED, BD, CE*) and the pair of points

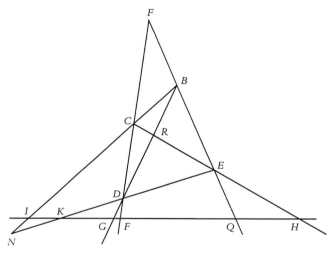

**Fig. 8.22** Diagram for Desargues' Involution Theorem. The vertices of the complete quadrilateral (marker posts) are *BCDE*. The theorem states that the points *I, K; G, H* and *P, Q* are in involution.

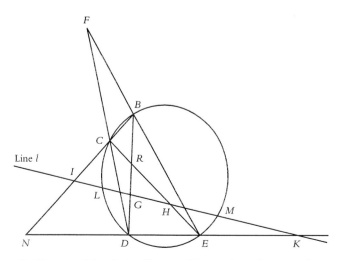

**Fig. 8.23** Second diagram for Desargues' Involution Theorem. The vertices of the complete quadrilateral (marker posts) are *BCDE*. The theorem states that the points *I, K; G, H* and *L, M* are in involution.

in which it meets the conic *BCDE*, the points *L* and *M*. By considering the other two possible pairs of sides of the quadrilateral, we obtain another two involutions like the one shown in Fig. 8.23.

Desargues' proof of his Involution Theorem is long, but rather elegant. It makes repeated use of Menelaus' Theorem (whose relevance is obvious from a comparison of Figs 8.22 and 8.16) and also uses the results about four points in involution deduced in the immediately preceding part of Desargues' work.

The Involution Theorem is important because it links conic sections with involutions. Desargues is taking it as obvious that all the conic sections can be projected one into another. The 'cone of vision' of the perspective books has become the cone of which the curves are sections—with the proviso that the eye looks in both directions at once, so as to give us the true cone, to either side of its vertex, as in the work of Apollonius. Now, Desargues has already proved that being six points in involution is a property that is not changed by projection. So any theorem concerning a conic that can be proved by considering involutions will be true of *all* conics. Projection preserves the involutions, so the proof will still hold, and we can arrange for the projection to give us any conic we please. This means that we can prove our theorem for the simplest conic, for instance for a circle, and then luxuriate in having proved a set of corresponding theorems for all the others. One does, of course, have to watch carefully what will happen to particular points and lines as the projection is carried out (especially if certain points and lines are projected to infinity) but essentially Desargues has provided a powerful way of tidying up the theory of conics. He has shown certain theorems are not merely *analogous*, that is expressible in terms which alter in more or less predictable ways from case to case, but are in fact strictly *equivalent*, that is capable of being proved by a single proof. This is, as it were, the positive side of his otherwise rather infuriating habit of trying to get as much as possible into one proof.

## Mathematics for mathematicians

One can sum up Desargues' achievement by saying that he has unified the theory of conics. He has done this by putting the conics back in their cone instead of considering each as a separate plane curve as Apollonius did. In fact, all ancient geometers prefer to reduce their problems to the plane—a habit that certainly makes it easier to draw the diagrams. We have seen this ancient style in the work of Commandino. It can be found also in the many Renaissance attempts to 'restore', that is, to reconstruct, the missing books of Apollonius' *Conics*, of which we shall have more to say below. Desargues, like Giovanni Battista Benedetti and Commandino's pupil Guidobaldo del Monte, attacks his problems in a three-dimensional way that, in competent hands, is almost bound to lead to results unknown to the ancients. The 'almost' is, however, significant because whereas Benedetti and Guidobaldo do come up with new results, what Desargues comes up with is a new method. He is, as it were, the mirror image of Benedetti. The latter arrived at the results he needed and apparently saw no particular significance in the method he had used. Desargues invents dozens of new names, apparently at least partly to draw attention to the newness of his method, but does not actually prove any theorems that are not to be found in either Pappus or Apollonius. From the twentieth-century viewpoint, Desargues' apparently exclusive concern with method seems much less frustrating than Benedetti's concentration on results. However, the 'look behind you!' style of history writing becomes tempting yet again when we look at the reception of Desargues' work by his contemporaries.

Desargues' *Rough draft on conics* is not a simple book, and much of what it has to say is new, though the style of exposition is generally conventional. Desargues' text perhaps seems harder going because one is not used to encountering this kind of repetitive intricacy in the vernacular. The otherwise highly comparable passages of argument in Commandino are all in Latin. As it happens, we do not have to ask ourselves to whom Desargues was addressing himself: it is known that the *Rough draft on conics* was printed for circulation to mathematicians, essentially to Marin Mersenne's extensive circle of friends and correspondents. As this readership was international—for instance, the recipients of copies of the work included the young English mathematician John Pell (1611–1685)—it is clearly of some significance that Desargues chose to write in French rather than in Latin. He is presumably allying himself with the mathematics of the practical tradition (or at least intending to distance himself from the university one). Desargues was not alone in his time in using the vernacular for works addressed to the learned. Famous examples from the late 1630s include Galileo Galilei's *Discourses and mathematical demonstrations concerning two new sciences …* (*Discorsi e dimostrazioni matematiche intorno à due nuove scienze …*, Leiden, 1638) and René Descartes' *Geometry* (*Geometrie*, Paris, 1637).

Unlike Descartes, however, Desargues had also written works that were in principle addressed to practitioners, including his brief treatise on perspective. The reaction of mathematicians to his work on conics in fact shows certain parallels to the reception of his work on perspective by artists. It seems possible, indeed, that it was coloured by it, or at least that some similar factors were at work. So in the next chapter we shall look again at the practical world of art before turning attention to mathematicians' responses to Desargues' *Rough draft on conics*.

# Chapter 9

## FRAGMENTED PERSPECTIVES

By the middle of the sixteenth century it was regarded as normal for an artist to have sufficient mathematical skill to give an appearance of three-dimensionality to pictures, as and when this was required. Later in the century there were in fact some artists who specialised in going much further than this. They painted illusionistic pictures in which at least some elements were to be seen as part of the real architecture of the room. These pictures make very bad illustrations in a book, because once everything is a photograph one truly has no way of telling whether the ornate carved frame shown round a picture is real or merely painted. Moreover, the change of scale also usually affects the strength of the illusion. A lesser or somehow more 'normal' degree of uncertainty seems to be required for this kind of sophisticated *trompe l'œil* to be fun to look at.

### Pictures and treatises

Small extremely naturalistic pictures, usually in the form of architectural views, real or imaginary, do not present the same kind of ambiguity. Scale tells one that this is a picture, and the artist's skill is to be judged by his success in persuading us that the picture is realistic rather than in convincing us that what he shows is actually present in our space. This kind of small picture was particularly popular in northern countries, where it can be seen as continuing a long-established tradition of naturalistic painting of landscapes. (The influence of this northern style can be seen in the detailed landscapes in the backgrounds of some Italian Renaissance pictures, for instance, in Piero della Francesca's portraits of Federigo da Montefeltro and Battista Sforza, Fig. 4.13.)

Perspective treatises concentrate attention on what can be done mathematically, and consequently, as we have seen, tend to be much concerned with architecture. This is ideally useful for the painter of detailed naturalistic townscapes. In fact, such pictures sometimes bear a disconcertingly close resemblance to plates in perspective treatises (see Figs 9.1 and 9.2). It is, of course, possible that the artist was specifically commissioned to base his picture on the finely engraved plates in one of the luxuriously illustrated treatises that appeared in this period. Such treatises certainly seem to be intended to appeal to picture collectors, and they give remarkably little guidance to an artist, except by providing architectural schemes that can be copied directly. Presumably the artist was expected to know the rudiments of the craft already, and thus be able to manage to make useful sense of the rather sketchy instructions provided in the book.

The picture by Dirck van Delen shown in Fig. 9.1 is almost certainly not closely based on a real scene. However, some of the artist's contemporaries did paint pictures that were intended as accurate renderings of real architecture, in effect portraits of buildings. The greatest of these architectural portraitists was Pieter Saenredam (1597–1665), who was particularly known for his pictures of the interiors of churches. His portrait, done by a friend Jacob van Campen (1595–1697)—a very

**Fig. 9.1** Dirck van Delen (1605/7–1671), *An Architectural Fantasy*, oil on panel, 46.7 × 60.5 cm, signed and dated 1634, National Gallery, London.

successful architect as well as a painter—is shown in Fig. 9.3. Saenredam's method of working was to begin by making a freehand sketch of the place concerned. These sketches are detailed, and usually annotated and dated, but they are also wonderfully atmospheric, particularly in giving a subtle account of the fall of light. Saenredam marks the point in the picture that is 'opposite' his eye, that is the foot of the perpendicular from the eye to the picture plane, the point which will be the 'centric point' when he comes to draw a picture in formal perspective. He calls this point the 'eye point', thus linking it to the observer, rather than merely to the construction. It was sometimes years later that Saenredam went on to the next stage, that of making an accurate perspective drawing. However, these drawings also are far from dry, though they are neatly marked with numerous orthogonal and transversal construction lines, and a centric point, and are usually made the same size as the final picture, so that they can be transferred and used as a basis for the finished painting. In these mathematical drawings we again find a delicate rendering of light, and the composition in the plane is perfectly balanced. In Saenredam's drawings and in his finished paintings, we have the mathematical skill that gives due account to scientific correctness, united with a sensi-

**Fig. 9.2**   Plate from Vredeman de Vries, *Perspective ...*, The Hague and Leiden, 1604–1605.

bility that gives us an art that hides art. As can be seen in Fig. 9.4 (despite the inevitable loss of quality when a picture is so reduced in size) Saendredam achieves an effect of light and calm that has its closest parallel in the work of Piero della Francesca. One cannot help feeling that, for both these painters, mathematical skill is so much part of their characters that it becomes an integral, though largely invisible, element in their art.

The architectural elements that make up almost all the picture in the case of townscapes, and paintings of the interiors of buildings, also provide the basis for the full scale illusionism of interior decorators or the perspectival schemes of large pictures such as altarpieces. So, mathematically speaking, the perspective treatises continue to look much the same as before, but the standard of their plates gets better, as copper engraving takes over from the woodcut.

## Abraham Bosse

It is within this world of competent engravers and an established tradition of mathematical, but not very mathematical, treatises on perspective that we must see the work of Abraham Bosse—to which we have already referred in Chapter 8. With hindsight—and historians' hindsight is always perfect—Bosse can be seen to be asking for trouble. As merely an engraver, he was not a full member of the Royal Academy of Painting and Sculpture (Académie Royale de Peinture et de Sculpture, founded 1648), whose president was the painter Charles Le Brun (1619–1690). As befitted an artist employed to express the dignity of the Court of Louis XIV, Le Brun's pictures

**Fig. 9.3** Jacob van Campen (1595–1657), Portrait of Pieter Saenredam, 1628, black chalk on paper, oval, 23.5 × 18 cm. Like Brunelleschi, Saenredam was a hunchback (Courtesy of the British Museum).

tend to be large and formal—to the point that in the twentieth century it requires a conscious effort to avoid describing them as 'pompous'. Their style was, however, greatly admired in their own time, and the contrast with the small and apparently much more informal works of Bosse would have been as evident then as it is now. As can be seen in Figs 9.5 and 9.6, even the loss of colour, and reduction of the works to approximately the same size, leaves much of the contrast intact. There was, as we have already implied, a comparable contrast in the social standing of the artists concerned. Twentieth-century viewers may prefer Bosse's evocations of scenes of the everyday life of the bourgeoisie to Le Brun's more visibly theatrical celebrations of the italianate classical style of the nobility, and one may even congratulate oneself on the twinge of democratic conscience that leads one to side with the underdog, but there can be little doubt who was the underdog. Bosse lost. That is to say, the conclusion of the curious artistic cum mathematical rumpus that followed Bosse's introduction of Desargues' perspective method into his teaching at the Academy was that Bosse lost his post as a teacher.

Personal frictions no doubt played their part in all this. Perhaps we are seeing some effects of the ill will that must surely have been stirred up by the long-running court case in which Desargues attempted to recover the very large sum of money he claimed was owing to his family. The purely

**Fig. 9.4**  Pieter Saenredam (1597–1665), interior of the Great Church at Haarlem, oil on wood, 59.5 × 81.7 cm, National Gallery, London.

intellectual grounds for the quarrel over perspective methods seem slender even by the standards of seventeenth-century learned controversies. Publishers and booksellers encouraged such disputes, since the flow and counterflow of argument sold books, but not all combatants enjoyed them: in the 1670s Isaac Newton (1642–1727), caught up in a controversy about some of his work on light, complained that Philosophy (that is, natural philosophy) was 'a litigious lady', whom he proposed to abandon. Newton could not in fact hold back from controversy, but he did arrange that his long, acrimonious and wide-ranging dispute with Leibniz (1646–1716) over the calculus was carried on through intermediaries—though Newton himself made extensive contributions to many of the writings that went out under other names. Desargues apparently shared Newton's distaste for personal combat. Having no professional mathematical credentials to defend, he seems merely to have stood back and let others fight for him. He could, of course, afford to pay for pamphlets to be published in his defence, and in that of Abraham Bosse.

The ill will began as a priority dispute: in 1643 there appeared in print a book on perspective which used a method of construction close to that described by Desargues. The title-page stated that the treatise had been written by the late Jacques Alleaume (1562–1627). Desargues was accused of having had access to Alleaume's papers and of having used them in his own work, published in 1636 and 1639. That is, the charge of plagiarism applied not only to the perspective

**Fig. 9.5**   Charles Le Brun, oil on canvas, *The Apotheosis of Louis XIV*.

construction, the matter that exercised the minds of the Royal Academy of Painting and Sculpture, but also to the new geometrical method in Desargues' book about conic sections. Comparison of the works concerned makes the latter part of the charge seem implausible. Alleaume's work (if it was indeed his) is fairly elegant but rather lightweight. It could perhaps have been derived from the work of Benedetti, picking up some of the methodological conclusions about projection that, to a twentieth-century eye, seem to be staring one in the face from the pages of the appendix to his book on sundials of 1574 (see Chapter 8). What Alleaume's book lacks is what Desargues so strikingly possesses: a tough-minded concern with the method of projection in its own right. As we shall see, this quality in Desargues is not in fact well represented even in the friendly reactions to his work. That the volume by Alleaume has this much in common with contemporary documents known to be reactions to Desargues' work tends to strengthen the suspicion that we are dealing with a text that has at least been subjected to some post-1639 manipulation. The late mathematician's manuscript would presumably have required editing before publication. It is, however, really the charge of plagiarism in regard to the method of perspective construction that wings the historian's mind with thoughts of malice against Desargues on the part of Alleaume's editor.

   This latter charge was, indeed, the one which received most attention at the time. The rights and wrongs of the matter are unlikely ever to be decided—if only because as a historical problem they are utterly uninteresting. In this period, there were several new perspective methods which, like that given in Desargues' perspective treatise of 1636, avoid the use of points

**Fig. 9.6**   Abraham Bosse, engraving, *The Visit*.

lying outside the picture field. The practical usefulnes of such 'abbreviated' constructions is obvious. Their intellectual inspiration is very probably to be found in Guidobaldo del Monte's widely read *Six books on perspective* of 1600, which, as we have seen in Chapter 7, contains a series of theorems that add up to a proof that any set of parallels in the scene to be portrayed will appear in perspective as a set of lines converging to a point. Later perspective treatises addressed to artists indeed show a rather tedious degree of interest in this result, which appears to be a source of the designation of various constructions as 'single point perspective', 'two point perspective', and so on—a nomenclature that persists to this day and is liable to be a considerable barrier to understanding pre-1600 constructions. It appears that Desargues and Alleaume were among the first to publish an 'abbreviated' perspective construction, but that several similar constructions appeared shortly thereafter. The evidence for some degree of simultaneous invention is rather strong. Moreover, the mathematically interesting part of the priority certainly rests with Guidobaldo del Monte. We know that historically this one ran and ran, but on closer inspection it has very little intellectual stamina.

A tacit recognition of this weakness may perhaps be detected behind the fact that another objection was also raised against Bosse: that his mathematical perspective was in any case not in accord with the realities of natural vision. Theories of vision had, by this time, changed in a way that did indeed make it impossible to claim (as Piero della Francesca had done in the fifteenth century) that perspective was a legitimate mathematical extension of the geometrical optics of sight. The nub of Le Brun's objection to perspective was thus simply the recognition that the two eyes do not act exactly as one but provide two slightly differing images of the outside world, which our mind then fuses into a stereoscopic view. This suggestion had been put forward, with a characteristic absence of ifs and buts, in René Descartes' *Optics* (*L'Optique*, Leiden, 1637). It was met with an assent that would seem entirely explicable were it not that natural philosophers had, without apparent unease, believed something very different for centuries past. A similarly rapid assent had greeted Kepler's proof that the picture formed at the back of the eye (on the retina) is upside down. This result was first printed in Kepler's treatise on optics of 1604, but was given wider circulation by appearing in Descartes' work in 1637. The clue may be that in neither case was there any important therapeutic consequence to be drawn from the new description of the functioning of the eye. So physicians, who were those with the greatest claim to expertise on the subject, could merely accept the new ideas, and carry on treating patients in the same way as before.

The above rational explanation of intellectual components of the battle over perspective in the Academy cannot explain why the battle lasted so long or why it was so bitter. Nor do they explain why, curiously enough, both sides in the combat made use of arguments based on a treatise that was far from modern, namely Leonardo da Vinci's *Treatise on painting* (*Trattato della pittura*), which first found its way into print, in Italian and in French translation, in Paris in 1651. Leonardo's book, which is an orderly compilation from surviving fragments in his manuscripts rather than a finished treatise, has very little to say about mathematical perspective. According to Vasari (writing about 35 years after Leonardo's death), Leonardo had abandoned the idea of writing on the mathematical part of painting when he learned that Piero della Francesca had already written about it. Unlike Piero's work, Leonardo's treatise contains a lot of painterly and rule-of-thumb advice. Mathematics appears mainly in the form of numbers used to express proportions. As its use by both sets of combatants in the Academy showed, Leonardo's treatise could be presented either as the authoritative antithetical alternative to Desargues' mathematical perspective or as the perfect artistic complement to it. The two sides were agreed that Leonardo was a genius. However, in the heat of battle, the anti-Bosse faction even went so far as to say that artists had nothing to learn from a mere mathematician. Leonardo would not have recognised that sentiment. The times had indeed greatly changed since the 1430s, when Alberti had attempted to claim for painters something of the intellectual and social respectability of mathematicians.

## Laurent de la Hyre

One of Bosse's more prominent defenders in the Academy was Laurent de la Hyre, a painter less eminent than Le Brun, but a man of some considerable reputation. In 1649 and 1650 Laurent de la Hyre painted a series of canvases showing the seven Liberal Arts (that is, the arts of the medieval university curriculum: the *trivium* of Grammar, Rhetoric and Dialectic, and the *quadrivium* of Arithmetic, Geometry, Music and Astronomy). The subject, which is rather old-fashioned for its time, was presumably prescribed by the patron who commissioned the pictures. They were

**Fig. 9.7**   Laurent de la Hyre, *Grammar*, oil on canvas, 102.9 × 113 cm, National Gallery, London.

intended as decoration for a room in a very large private house, belonging to a M. de Tallement, but the surviving canvases seem to belong to two different series, one of a wide rectangular format, the other more nearly square. The less wide pictures are, in effect, copies of the central parts of the wider ones, though there is no good reason to suppose that the less wide pictures were not painted at the same time as the ones from which they appear to be extracts. Over the years, the sets of seven canvases have been dispersed (Figure 9.7 shows the nearly square version of *Grammar* that is now in London), and the wider rectangular version of *Geometry* emerged from a private collection only in 1993. The two versions of *Geometry* are shown in Figs 9.8 and 9.9. The part on the left in the wider version (a part that was excised in the other version) is of considerable historical interest because it includes a representation of a perspective construction that is almost certainly intended as a reference to the method of Desargues. The lines that show this construction in La Hyre's painting have apparently been drawn in pure white, probably to make them stand out. However, the lines are very fine, and the intensity of the white has been somewhat weakened by the ageing of the pigment, so the lines do not show up well in photographs. A redrawn version of them is shown in Fig. 9.10. As can be seen by comparison with the diagrams from Desargues' perspective treatise (Figs 8.13 and 8.14), the lines shown by La Hyre are not exactly what would be required to produce the picture he shows on the easel—and it would, in any case, hardly be realistic to show

**Fig. 9.8**   Laurent de La Hyre, *Geometry*, 1649, oil on canvas, 106.1 × 158.7 cm, Ohio, Toledo Museum.

**Fig. 9.9**   Laurent de La Hyre, *Geometry*, 1649, oil on canvas, 104 × 218 cm, Private collection.

them on top of the final painting. However, the pattern of lines shown is very like patterns obtained by using Desargues' construction—indeed it may have been copied from one of the diagrams in Bosse's expanded version of Desargues' treatise, published in 1648 (which will be discussed below). In any case, the pattern La Hyre shows is not at all like the pattern that would appear from using conventional perspective constructions.

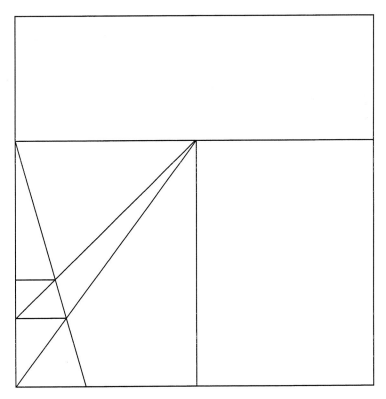

**Fig. 9.10**  Drawing after a detail of Laurent de La Hyre's *Geometry* (Fig. 9.9) showing the construction lines on the painting displayed on the easel at the left. Copy of a drawing by M. J. Kemp.

Further evidence of La Hyre's interest in the work of Desargues is to be found in the diagrams shown on the sheet of paper Geometry is brandishing for our attention. The sheet of paper is shown separately in Fig. 9.11. The diagram at the top left is Euclid's figure for his proof of Pythagoras' Theorem (*Elements* 1, Proposition 47). This diagram would certainly have been easily recognised by an educated person of La Hyre's time. In fact it is a standard emblem used to identify personifications of Geometry on decorative frontispieces of mathematical books. The significance of the other diagrams on Geometry's piece of paper is not quite so obvious. However, it seems reasonable to suppose that they too come from the *Elements*. Since Euclid is extremely systematic, and does not repeat himself, any particular diagram belongs to one particular theorem in the *Elements*. Therefore, unlikely as this may at first seem, it is in fact possible to associate each of the diagrams La Hyre has shown with precisely one of Euclid's propositions. (For the record: the diagram at the right is for *Elements* 2, Proposition 9, and that at bottom left is for the first case of *Elements* 3, Proposition 36.) It turns out that these two propositions, and Pythagoras' Theorem, are all results which, in his *Rough draft on conics*, Desargues proposes to generalise. That is, the three theorems stated by Euclid would become special cases of three more general theorems proposed by Desargues. There are in fact only four propositions of Euclid which Desargues proposes to generalise in this way. The fourth has no identifiable diagram since it refers only to line segments—which may explain the array of lines shown on the far right of the sheet of paper in Fig. 9.11. All in all, the collection of diagrams La Hyre has shown could have been selected by Desargues

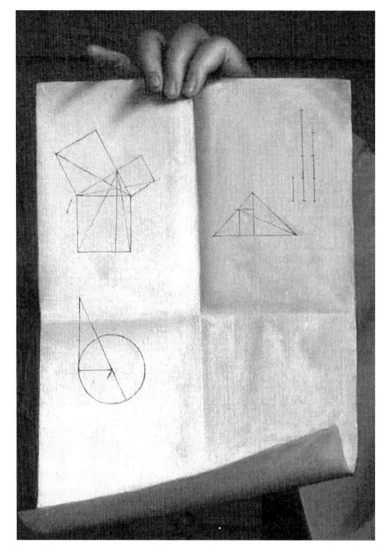

**Fig. 9.11**  Detail of Fig. 9.9, showing the sheet of diagrams. The diagrams are all from Euclid's *Elements*. The diagram at the top left is for Book 1, Proposition 47 (Pythagoras' Theorem), that at the right is for Book 2, Proposition 9, and that at the bottom left is for the first case of Book 3, Proposition 36, a result Desargues uses repeatedly in his *Rough draft on conics*.

himself. Perhaps La Hyre actually asked him which of the results in Euclid were of particular interest to a modern geometer?

La Hyre's *Geometry* also makes a much simpler point in the perspective battle. Most unusually for a picture of this date, it provides the information we require to find the centric point of its perspective. Moreover, this compositional manœuvre must surely be deliberate, because La Hyre has in fact supplied the exact minimum that is needed. The two lines at top and bottom of the canvas displayed on the easel represent a pair of parallel lines that are also parallel to the horizon. Therefore, by a theorem proved in Guidobaldo's treatise of 1600, the lines in the painting meet at

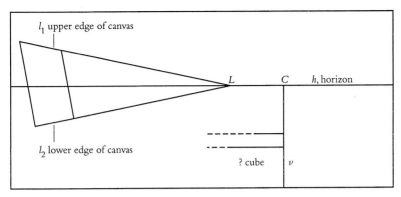

**Fig. 9.12** Drawing based on a tracing of a photograph of La Hyre's *Geometry* (Fig. 9.9) to show the perspective scheme of the complete painting. Lines $l_1$, $l_2$ represent the upper and lower edges of the canvas displayed on the easel at the left. Line $h$ is the horizon of the ideal viewer. Line $v$ represents the right face of the solid on which the figure of Geometry is leaning. The centric point of the perspective is at $C$.

a point on the horizon of the main picture. In Fig. 9.12, which is based on a tracing of a photograph of the painting shown in Fig. 9.9, the two horizontal edges of the canvas on the easel are the lines $l_1$ and $l_2$. Their meeting point $L$ defines the level of the horizon in the complete picture, the line we have called $h$. This line marks the level of the eye of the ideal viewer, and the centric point lies somewhere on it. We can find exactly where by looking at the stone shape on which Geometry herself is leaning. This solid is probably meant to be a cube—to go with the cube shown in the canvas displayed on the easel to the left. In any case, it is obviously meant to have faces that are vertical and horizontal, so the fact that on the right we see the (vertical) right edge of the front face exactly aligned with the (horizontal) right edge of the top face, and aligned so as to give a vertical in the picture, shows that the eye of the ideal viewer must lie in the vertical plane defined by these two edges (the vertical line we see being the intersection of this plane with the plane of the picture). La Hyre has emphasised the compositional significance of this line by providing a sort of visual rhyme to it at the right edge of the plinth supporting the sphinx. The right side of the outline of the plinth is not quite straight: the line marking the edge of the upper surface is angled slightly inwards, and as if to draw our attention to this break there is also a break in the vertical edge, in the form of a chip taken out of the stone of the plinth. In Fig. 9.12 the line representing the right face of the solid on which Geometry is leaning has been called $v$. The intersection of the lines $h$ and $v$ gives us the centric point of the perspective, the point called $C$ in Fig. 9.12. This is as far as we can go: we are not given any information that would allow us to find the viewing distance and thus the ideal viewing point for the picture.

Like the main picture, the canvas displayed on the easel at the left also provides us with the information required to find the centric point of its perspective—and, indeed, a viewing distance, if we are willing to assume a shape for the geometrical solids it shows. However, this is very much a matter of principle. In practice, the lines we are given are short, making actual reconstruction a decidedly inaccurate procedure. We can use pairs of images of horizontals, such as the top and bottom edges of faces of the geometrical solids, to find points on the horizon of the picture. This horizon should, in reality, be parallel to the top and bottom edges of the canvas, so the images of all three lines will meet in a single point on the horizon of the main picture. In the painting, the lines at the top and bottom of the canvas are those we have called $l_1$ and $l_2$ in Fig. 9.12. They meet at $L$, so the horizon of the picture within the picture should also pass through $L$. La Hyre presumably constructed it to do so. In fact constructing the perspective of the picture on the easel is not so difficult a matter as it may appear at first sight. The convergence properties we need to use are projective, so the sets of parallel lines in the picture can be constructed exactly as if it were to be seen straight on. Desargues' horizontal and vertical scales could be used to establish dimensions in very much the way he described in his treatise of 1636 (see Chapter 8).

In fact, in 1648 Abraham Bosse had published a substantial perspective treatise based on Desargues' work, and had included in it a detailed treatment of the problem of constructing a picture whose plane is not perpendicular to the line of sight. As we have seen in Chapter 7, this problem had already been treated by Giovanni Battista Benedetti in 1585, but his work is not in a style to make it useful to artists and there is no evidence that either Desargues or Bosse knew about it. In fact, Benedetti's work as a whole seems to have been fairly comprehensively overlooked by mathematicians of the next generation, and an embarrassingly large number of historians followed their example, with the result (among others) that credit for being the first to consider obliquely placed pictures is often given to Desargues and Bosse.

## Bosse's treatise of 1648 and Desargues' Perspective Theorem

Abraham Bosse's treatise on perspective is entitled *Universal manner of Mr Desargues for the pratice of perspective by the use of scales and plans …* (*Manière universelle de Mr Desargues pour pratiquer la perspective par petit pied, comme le geometral. …*, Paris, 1648). It was printed in octavo, a format which suggests that the volume was intended as a handbook for craftsmen. Theoretical works addressed to patrons were usually larger in size, that is folio or quarto. For instance, the edition of Leonardo's *Treatise on painting* printed in Paris in 1651—which, as we have seen, was used by both sides in the perspective battle in the Academy of Painting and Sculpture—is a quarto.

Bosse's style in his treatise on perspective is perhaps most kindly described as wordy. There is much detail, for which the intended readership was probably grateful. The practitioner was presumably expected to find the example that seemed closest to what he needed and to adapt it as necessary. That is, the book was to be used like all its practical predecessors. To a twentieth-century reader in search of mathematical enlightenment this makes for hard reading, though Bosse's skill as an engraver ensures that the elegance of the abundant illustrations makes some atonement for the style of the text. An echo of the perspective battle can be found in the fact that the main body of Bosse's treatise is followed (on page 321) by a reprint of Desargues' brief work on perspective of 1636. The text has been reset, but the differences from the original are so slight that one cannot really regard this as a second edition. Bosse's purpose in printing this text is almost certainly to establish Desargues' priority over Alleaume in regard to the method of construction it describes.

Desargues' text is followed by two propositions to do with perspective and painting whose discursive style indicates that they are by Bosse. The tone then abruptly changes back to that of Desargues for the final three propositions, each with a heading to say it is geometrical. The style of these propositions is like that of the *Rough draft on conics*, formal and terse, with proofs that proceed with the minimum of explanation.

The first of the three geometrical propositions is what is now known as Desargues' Theorem on triangles in perspective. Today this is usually stated in the form: if two triangles are in perspective,

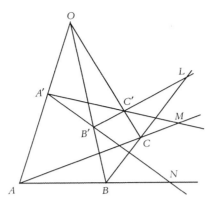

**Fig. 9.13** Diagram for Desargues' theorem on two triangles in perspective, in modern form. The triangles in perspective are *ABC* and *A'B'C'*, so that *AA'*, *BB'*, *CC'* meet at the centre of perspective, *O*. The meets of corresponding sides of the triangles are as follows: *BC*, *B'C'* meet in *L*; *CA*, *C'A'* in *M*; *AB*, *A'B'* in *N*. Desargues' Theorem states that *L*, *M*, *N* lie on a straight line.

then the meets of corresponding sides are collinear. This formulation is perhaps too neat to be clear. What is going on can be seen in Fig. 9.13. The two triangles are *ABC*, *A'B'C'*. Since the triangles are in perspective, the lines *AA'*, *BB'*, *CC'* meet at the centre of perspective *O*. Corresponding sides of the triangles meet as follows: *BC*, *B'C'* meet in *L*; *CA*, *C'A'* in *M*; *AB*, *A'B'* in *N*. Desargues' Theorem states that *L*, *M*, *N* lie on a straight line.

Desargues' own statement of the theorem is much less clear than the modern one we have just given, though some help is provided by Bosse's diagram (shown in our Fig. 9.14). Desargues' statement of his theorem goes

When the straight lines *HDa*, *HEb*, *cED*, *lga*, *lfb*, *Hlk*, *DgK*, *Efk*, which either lie in different planes or in the same one, cut one another in any order and at any angle in such points [as those implied by the lettering]; the points *c*, *f*, *g* lie on a straight line *cfg*.

He then launches into the proof, without so much as beginning a new paragraph. This is, indeed, the style of writing, and of reasoning, that we found in Desargues' *Rough draft on conics*. However, on this occasion, the user-unfriendly nature of the text has contributed an element of comedy. Readers self-confident enough to attempt to follow the mathematical sense of Desargues' statement (and the present author admits to having belonged to this group only through the necessity imposed by making a translation) will have noticed that the statement, as given, cannot be correct. We require two triangles, with six sides between them, plus the three lines through pairs of vertices that establish the triangles are in perspective, that is another three lines, making nine in all. The theorem then claims the existence of a tenth line, one through the meets of corresponding sides. Desargues' statement of his theorem, as printed in 1648, gives only eight lines, and then claims the existence of a ninth. The missing line is *cab*. It does appear in the accompanying diagram, so we presumably have a proofreading error rather than a mathematical howler by Desargues or Bosse. The original error was probably made by a mesmerised printer, since at this time authors did not usually read their own proofs. So far so understandable—though a bit conspicuously unfortunate. What are we to make, however, of the fact that the error is repeated in the text of the proposition given in the two standard editions of Desargues' works, published in 1864

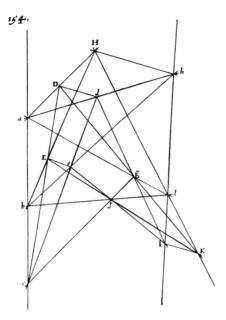

**Fig. 9.14**  Original figure for Desargues' theorem on two triangles in perspective, from Abraham Bosse, *Universal manner of Mr Desargues for the practice of perspective* … (*Manière universelle de Mr Desargues pour pratiquer la perspective …*, Paris, 1648), Plate 154 (opposite page 340).

and 1951 (the latter reprinted in 1981)? It is surely an eloquent tribute to the obscurity of Desargues' style that such a simple omission could occur in so important a passage without attracting the attention of the two highly competent mathematicians who edited his text. As misprint stories go, this one seems to have a moral.

As Desargues has noted in his statement of the theorem, it holds irrespective of whether the two triangles lie in the same plane. He effectively gives separate proofs for the two cases, but actually goes through both at the same time, alternating between cases as necessary. The proof for the case where the triangles do not lie in the same plane is very simple and elegant. That for the case where the triangles are coplanar depends upon repeated use of Menelaus' Theorem (the metric theorem about transversals which Desargues used in his work on conics—see Chapter 8). Desargues' combination of the two proofs is, characteristically, hard to follow.

It seems a pity to temper the simplicity of the proof of the first case by combining the two proofs or making use of Desargues' unhelpful notation. In what follows we shall therefore consider the two cases separately and our references will be to Fig. 9.13, in which the two triangles in perspective are $ABC$ and $A'B'C'$, their centre of perspective being the point $O$. Since any side of the triangle $ABC$, say $AB$, lies in the plane of the triangle, all the points on this line must lie in the plane of the triangle $ABC$. So the point $N$ in which $AB$ intersects $A'B'$ must also lie in the plane $ABC$. However, the line $A'B'$ lies in the plane $A'B'C'$, therefore all the points of the line $A'B'$ lie in the plane $A'B'C'$, therefore its point of intersection with the line $AB$, the point $N$, lies in the plane $A'B'C'$. So $N$, the meet of $AB$ and $A'B'$ lies in both planes $ABC$ and $A'B'C'$, therefore it must lie on the straight line in which they intersect. Similarly, it is clear that the meets of the other

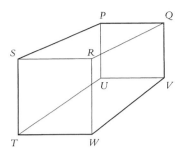

**Fig. 9.15** A cuboid, to show skew lines. The lines *QR*, *TW* are skew: they do not meet because they are not in the same plane (that is, it is not possible to construct a plane including them both).

pairs of corresponding sides, the points *L* and *M*, must also lie on this line of intersection. So the points *L, M, N* are collinear.

The above does not, however, amount to a complete proof of this case of the theorem: we have not actually proved that the lines *AB* and *A'B'* do intersect. This objection does not refer to the possibility that the lines are parallel, which, as we have seen, has been taken care of by Desargues' definition of points at infinity. The possibility that needs to be eliminated is that of the lines being, in mathematical parlance, 'skew'. This is a property peculiar to lines that are not in the same plane. (Once we take account of points at infinity, it can be seen that parallel lines are in the same plane.) Examples of skew lines can be seen in Fig. 9.15. For instance, the lines *PS* and *UV* do not meet, and it is clear that no amount of extending their lengths will make them do so. It is quite easy to see that one cannot construct a plane containing the line *SP* which will also contain the line *UV*. Once one has admitted Desargues' points at infinity, it becomes an axiom that if two lines lie in the same plane they will meet.

If we now go back to Desargues' Theorem, we can see that it is in order to establish that the corresponding sides of the triangles *ABC, A'B'C'* are coplanar, two by two, that the statement of the theorem imposes the condition that the two triangles are in perspective. For, since the two triangles *ABC, A'B'C'* are in perspective, with centre of perspective *O*, the lines *AA', BB', CC'* all pass through *O*. Therefore the lines *AA', BB'* define a plane. Since all four points *A, A', B, B'* must lie in this plane, it is clear that the lines *AB, A'B'* also lie in it. Therefore, since these lines are coplanar, they will meet.

Writing this proof out in detail takes rather a lot of words, but the essence of the matter is very simple. One merely has to see the planes in question and recognise their interrelationship. To give a formal proof would require reference to little more than the first few theorems in Book 11 of Euclid's *Elements*.

In the case in which the triangles *ABC, A'B'C'* lie in the same plane, we do not need to prove that the pairs of corresponding sides will meet. On the other hand, we cannot construct the line *LMN* as the intersection of planes, and a complelely different style of proof is required. Desargues' proof proceeds by repeated use of Menelaus' Theorem, that is by considering relations between the ratios of lengths of segments cut off on sides of triangles by transversal lines (see Chapter 8, especially Figs 8.16 and 8.17).

Desargues did not give a name to his new theorem. It is presented merely as a 'geometrical proposition'. The modern name, Desargues' Perspective Theorem, records its being published in a treatise on perspective and serves to distinguish it from the Involution Theorem, but is otherwise somewhat misleading. The theorem is best seen as being not about perspective but about projection. It is clear that the two triangles that are 'in perspective' are equivalent, so either could be considered as an image of the other. Moreover, the fact that the theorem holds both when the triangles are in different planes and when they are in the same one establishes that we are dealing with projective properties. As a schoolchild, the present author jibbed at the elaborate proof of the plane case and attempted to maintain that it was (really) proved by the fact that one could draw a plane diagram for the three-dimensional case. The answer (not given by the schoolteacher concerned at the time) is that actually drawing the diagram is not enough: one has to prove it can be done. And that puts one back to proving the plane case of the theorem. Thus, while the theorem is merely a curiosity for perspective, it is fundamental for projection. The two following 'geometrical propositions' also seem to refer to projection rather than to perspective problems. It thus seems likely that all three propositions are developments of the work of the *Rough draft on conics*, though it is not clear how they arose.

This absence of apparent interconnections among results obtained by projection may perhaps help to explain the later fate of Desargues' work, namely its dropping out of sight for nearly two centuries before the main results were rediscovered independently, Desargues' name being associated only with a text of which no printed copy was known to survive. With hindsight, one can see that the immediate response to Desargues' work gave some further hints to explain future neglect.

## Mathematical responses and another geometry

The battle over perspective methods at the Academy of Painting and Sculpture almost certainly coloured contemporary mathematical responses to Desargues' *Rough draft on conics* of 1639. All the same, reactions specifically directed to the work on conics were more cautious than those concerning the perspective treatise, and were eventually comparatively friendly. It was, however, pointed out almost at once that the problems Desargues had solved could also have been been solved by ancient methods, that is by the methods used by Apollonius of Perga in his *Conics*. This is true. Indeed, specimen solutions were produced to show that it was true. Unfortunately, mathematicians of the time were apparently inclined to pursue this line of reasoning to the point of drawing the conclusion that Desargues' method was not capable of proving or generating new theorems that were not to be found by the ancient method. This deduction is, of course, mildly illogical, but the conclusion was shown to be seriously misleading only in the nineteenth century, when a large body of work was done in pure geometry following the reinvention of the projective method by Gaspard Monge (1746–1818) and his followers Michel Chasles (1793–1880) and Jean Victor Poncelet (1788–1867).

The use of the post-1639 term 'pure geometry' is a pointer to another reason why Desargues' work did not receive the acclaim and development that hindsight judges as its due. By 1639, and increasingly thereafter, the kind of geometry associated with the ancients, and with Desargues, which essentially depends only upon visualising geometrical figures—what is now called 'pure geometry'—had a rival: 'algebraic geometry'. One can see traces of what would now be called algebraic methods in the work of some ancient geometers, notably Apollonius of Perga, but it was the rapid development of algebra in its own right in the sixteenth century that led to mathematicians making increased use of it in their geometrical work. Curiously enough, little historical research has been done on the immediate origins of a fully algebraic geometry. Perhaps this is partly because there is no disagreement about the fact that the work in which such geometry first appears in print is René Descartes' *Geometry* (*Geometrie* (*sic*, no accents), Leiden, 1637). There was general agreement at the time that this work looked very interesting. Moreover, Descartes had taken care to point out, and to demonstrate, that his new method could be used to solve problems which had defeated ancient geometers. Thus mathematicians' interest in Cartesian algebraic geometry flourished alongside a continuing interest in finding out what was contained in the 'lost' parts of the work of Apollonius (who was effectively Desargues' rival in the treatment of conic sections). At this time only the first four books of Apollonius' *Conics* were known. They had been published in 1566, in a Latin version made from the Greek text by Federico Commandino. The remaining books turned out to survive only in Arabic. They were published in 1710, in a Latin translation by Edmond Halley (*c.* 1656–1743). Halley is now best remembered as an astronomer. His edition of Apollonius was apparently intended to establish his specifically mathematical learned credentials as a newly appointed professor at Oxford. His comet became an unassailable claim to fame only when it returned as he had predicted, in 1758. By Halley's time, the temporary (but quite long) ascendancy of algebraic geometry over pure geometry was assured. The newly invented calculus was proving to be a very powerful mathematical tool, and its use favoured the development of Cartesian geometry throughout the eighteenth century.

That is the long-term story. In the short term, matters went equally peaceably. Descartes was one of those to whom Marin Mersenne sent a printed copy of Desargues' *Rough draft on conics* in 1639. We know this because Descartes sent Desargues some comments on the book. Descartes had apparently found it interesting, but far from easy to read. Indeed some part of what deconstructionists would call the 'undertext' of his letter is the message that he, Descartes, is not willing to undertake the rewriting that would be necessary before wider publication. Descartes seems to be choosing his words with great care. He suggests Desargues must decide what readers he is addressing: learned mathematicians or people who are interested in the subject without being experts. He suggests that Desargues' use of novel French terms instead of ones derived from Greek and Latin

might even serve to attract some people to read your work, as they read works on Heraldry, Hunting, Architecture etc., without any wish to become hunters or architects but only to learn to talk about them correctly.

Desargues' French terms had included new names for the ellipse, parabola and hyperbola: *defaillement* or *ovale*, *egalation*, and *outrepassement* or *excedement*. While any translator naturally welcomes the rare opportunity of indulging in the invention of neologisms, one must at the same time wonder whether the use of Desargues' vocabulary—any more than the Englished versions deficity, equalation, excedency—would count as talking about the subject correctly (though we may note that Desargues does actually supply the normal Apollonian names as well). The raised eyebrow is not necessarily lowered when Descartes tells Desargues that, for such a non-expert readership, he must explain

everything so fully, so clearly and so distinctly that these gentlemen, who cannot study a book without yawning and cannot exert their imagination to understand a proposition of Geometry, nor turn the page to look at the letters on a figure, will not find anything in your discourse which seems to them to be less easy of understanding than the description of an enchanted palace in a novel.

Descartes had not written his *Geometry* for this kind of reader. One doubts whether he really believed such rewriting was possible for the *Rough draft on conics*—though we may note that Bosse's *Perspective of Mr Desargues* (1648) is a rewrite of Desargues' brief tract of 1636, in a style that approximates to Descartes' prescription, at least in being wordy and heavily illustrated. As we have seen in Chapter 6, perspective was, like hunting, heraldry and architecture, a subject one might wish to be able to talk about correctly, without becoming an expert on its technicalities. Perhaps the mind should not boggle at the thought that such conversation could stray onto the related topic of Desargues' new geometry. After all, La Hyre's painting of Geometry (Fig. 9.9) seems designed to contribute to just such discussion. In fact, we have here a connection with the next generation of mathematicians, through Laurent de La Hyre's son Philippe (see below), but we must first look at the work of the only contemporary mathematician who picked up Desargues' ideas and used them for himself, namely Blaise Pascal (1623–1662).

## Blaise Pascal

Pascal was in many respects a one-off. He has a deservedly high reputation as a mathematician, a philosopher, a religious polemicist and a natural philosopher. In the last capacity he made important experimental investigations of the properties of fluids, including a study of atmospheric pressure which involved the considerable exploit of carrying barometric apparatus up the highest accessible hill he could find, the Puy de Dôme (height 1465 metres above sea level, and near the town of Clermont-Ferrand). Almost all of Pascal's mathematics belongs to his youth. He later decided that his love for mathematics was drawing him away from his love for God, gave the subject up, and in 1654 retired to the convent of Port Royal. Unlike Desargues, Pascal writes with beautiful clarity. One might in fact call it passionate clarity. His short essay *The spirit of geometry* (*L'Esprit de la Géométrie*, Paris, 1657, probably written in 1654) reads like a declaration of love in a tragedy by Racine (1639–1699). (Racine, who was an orphan, was in fact brought up in Port Royal.)

While still in his teens, Blaise Pascal accompanied his father Étienne (1588–1651) to gatherings of the circle of mathematically-minded friends centred on Marin Mersenne. He must have known about Desargues' *Rough draft on conics* as soon as it appeared. His own response, *Essay on conics* (*Essay pour les coniques*), is dated 1640, but seems to have been written in 1639. It is notable for using some of Desargues' terms, for instance referring to an 'ordinance of lines' (the modern name is 'pencil'), and for employing Desargues' methods. It contains what is now known as Pascal's theorem, which is, in modern terms: if six points $A$, $B$, $C$, $D$, $E$, $F$ lie on a conic, and $AB$, $DE$ are produced to meet at $L$, $GC$ and $EF$ to meet at $M$, $CD$ and $FA$ to meet at $N$, then $L$, $M$, $N$ lie on a line (see Fig. 9.16). Pascal does not give a full proof of this result, but since he starts with six points on a circle it is clear that he proposes to obtain the six points on a general conic by projection from this diagram.

Pascal also states some further theorems, and it is known that he later developed some of the ideas he put forward in the *Essay*. However, the expanded text, which was called *Treatise on conics* (*Traité des coniques*), remained unpublished at Pascal's death and is now lost. We know a little about its contents from some notes made by Leibniz in 1676. Pascal's work can now be seen to be mathematically important, but it did not influence the work of the following generation. In fact, the loss of Pascal's treatise should really be seen as an indication that Desargues' work had failed to enter the mainstream mathematical tradition. In historical terms, Pascal's work failed to transmit

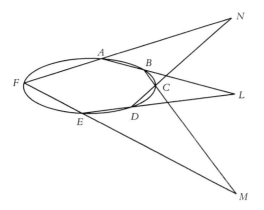

**Fig. 9.16** Pascal's Theorem.

Desargues' ideas—except, perhaps, in one respect. That is in its notion of space. It is possibly under the influence of Desargues' rigorous mathematical treatment of infinity in geometry that Pascal was able to accept the idea of infinite space as an entity in its own right. (The famous remark about being frightened by the silence of infinite space is not presented as expressing Pascal's own feelings but those of a postulated *libertin*, that is a 'free thinker' or atheist.) For Pascal explicitly, as by implication for Desargues, geometry has become the science of space. Space is now something in its own right, independent of matter and of measurement. That is the notion of space used by Newton (though there is no evidence that he derived it from Pascal).

## Philippe de La Hire

Consistent spelling of personal names is not a characterisitic of the seventeenth century. Confusion can usually be avoided by adopting the spellings favoured by the owners of the names. Philippe de La Hire (1640–1718), now best remembered as a mathematician, signs his writings with an 'i' in his surname, so historians of science spell it that way too. However, his father, Laurent de la Hyre, usually signs his pictures with a 'y' in the surname, and historians of art accordingly copy his example. The present author apologises for any confusion that may result from the divergence.

As we have already partly seen, there is not necessarily any great divergence in the intellectual interests of the two men. There are many features of Laurent's *Geometry* (Fig. 9.9) that suggest he had a serious interest in the learned discipline it sets out to portray. La Hyre's compositions are not in general notable for elaborate or detailed perspective, so there is reason to suppose that what we are seeing in this particular picture is the tip of an iceberg of otherwise invisible concern with mathematics for its own sake, rather than a mere by-product of a professional concern with mathematical correctness in pictures. Laurent's son Philippe was nine years old when his father painted *Geometry*, so we may imagine him taking an intelligent interest in it, and not only for its mathematical content. Philippe was to be trained as a painter in his father's workshop. One of his earliest writings was a treatise on painting, though the work was not published until 1780, when the Academy of Sciences (Académie des Sciences, founded in 1666) decided to print some papers dating from its first thirty years. In fact, Philippe de La Hire's writings show him as very much in the mould of the practical mathematicians with whom we have been concerned in earlier chapters. The activities of craftsmen seem to have been at the root of much of his mathematical work. For instance, his work on roulettes (the curves traced out by a point on the rim of a wheel rolling on another wheel or on a line) is apparently connected with problems involving gearing. It comes as no surprise to find that Philippe de La Hire's advice was sought on the repair of the wheelwork of a pump originally made to a design by Desargues.

Philippe de La Hire's mathematics also provides a piece of circumstantial evidence regarding his father's relationship with Desargues. In 1679, Philippe made a manuscript copy of Desargues' *Rough draft on conics*—wishing, as he said, to get to know the work better. It seems highly likely that the printed copy from which he worked had belonged to his father. To judge by the page numbers added to it, the copy in question is the one now in the Bibliothèque nationale in Paris. Until this copy came to light, in 1951, Philippe de La Hire's manuscript provided the only known text of Desargues' work.

La Hire's comment about wishing to get to know the *Rough draft on conics* better comes from the note he appended to his manuscript copy. We also learn from this note that La Hire had written his first book on geometry before studying Desargues:

In the month of July of the year 1679 I first read this little book by M. Desargues, and copied it out so as to get to know it better. It was more than six years since I had published my first work on conic sections. And I do not doubt that, if I had known anything of this treatise, I should not have discovered the method I employed, for I should never have believed it possible to find any simpler procedure which was also general in its application.

The 'simpler procedure' which La Hire had used in his *New method of geometry for sections of conic and cylindrical surfaces (Nouvelle méthode en géométrie pour les sections des superficies coniques, et cylindriques etc.*, Paris, 1673) had involved considering cross ratios of four points (specifically harmonic ranges, see Chapter 8 and especially equation (3)). As we explained in Chapter 8, Desargues had preferred to stick with the more general notion of six points in involution. However, La Hire's work resembled Desargues' in the important matter of treating problems by means of projection. The 'harmonic range', defined by its cross ratio being unity, is in fact a simplified form of involution, and shares with the involution the property of being invariant under projection. To a reader in quest of simplicity, or its natural companion clarity, it might seem that La Hire should merely proceed to congratulate himself on having done better than Desargues. That was not what he did. He continued and extended his work on conics, and eventually produced an extremely elegant treatise, called *Conic sections (Sectiones conicae*, Paris 1685), which incorporates the results of Desargues and Pascal, providing proofs where they had been lacking in the originals.

La Hire's *Conic sections* is thoroughly projective. But its method is not exactly modelled on Desargues'. La Hire's continued work on conics had presumably convinced him that nothing was lost, in dealing with conics, in using his simpler harmonic ranges of four points (derived, as he said, from Pappus and Apollonius) rather than Desargues' more general involution of six points. So in a way he did decide he had done better than Desargues. Certainly he had done better in regard to clarity of style. However, it seems that the visibly classical origins of his projective method tended to confirm his readers in their belief that all this was in Apollonius—in principle if not in historical fact. Since he was writing in Latin, La Hire was explicitly addressing himself to the international educated readership that could be expected to make comparisons between his work and that of Apollonius.

With hindsight, La Hire can be seen as having done slightly too good a job of naturalising the projective method into the learned tradition. Very much in the style typical of Renaissance scholarship, he had shown new work as a continuation of an ancient tradition. In doing so, he had taken from Desargues not his method of using involutions but the results obtained by it. (Effectively, he had done by Desargues' work rather as Giovanni Battista Benedetti had done by his own—which is not a good augury!) Moreover, La Hire had sacrificed something of the generality of Desargues' method.

## The next generation and beyond

As we have seen, the increasing popularity of calculus played an important part in consigning pure geometry to mathematicians' mental lumber room for about a century. With it went projective geometry. It is therefore somewhat ironic that both of the two contenders for the title of inventor of the calculus, Newton and Leibniz, did actually show an interest in projective geometry. We have already remarked on Leibniz's having made notes on the unpublished *Treatise on conic sections* by Pascal. Newton, in his *Mathematical principles of natural philosophy* (*Philosophiæ naturalis principia mathematica*, London, 1687), uses projection to find a conic that passes through a certain number of points and touches a certain number of lines. This is in the first book of the work; the famous inverse square law of gravity appears only in the third. On the whole, Newton's readers, who would have needed to be competent mathematicians if they were to get anywhere with the work, somewhat understandably looked to the *Principia* for the System of the World in the third book rather than the mathematics in the first. Newton himself apparently knew about projection from the work of La Hire. One can only speculate on what might have been had he come across a copy of Desargues' original text.

One of the combatants for Newton in the continuation of the calculus battle was Brook Taylor (after whom a theorem in calculus is named). Showing another side of his mathematical character, Taylor (1685–1731) also wrote on perspective. In fact it was his treatise *Linear perspective* (London, 1715) that introduced the term that forms its title, and the name 'vanishing point' that is now used to denote what Renaissance texts call the 'centric point' or (less understandably) the 'eye point'. Taylor does not quite explain what is supposed to vanish at the vanishing point, but the name has stuck. The readers he was addressing were presumably not expected to be so literal minded as to ask that question. The book, a mere 45 octavo pages, with eighteen plates, is apparently addressed to craftsmen. It is about how to draw, and some results are given without proof. However, Taylor seems to require his readers to be competent mathematicians, and even considers the problem of finding the true dimensions of an object from a perspective drawing. This is a problem that seems more suited to appeal to a patron of architects than to a draughtsman, unless he was unusually good at mathematics. Perhaps Taylor was mindful that his readers might include Sir Christopher Wren? Before turning to architecture, Wren had been a professor of astronomy at Oxford. He was a very skilled mathematician. In any case, the second edition of Brook Taylor's treatise was given a slightly less dry title, *New principles of linear perspective* (London, 1719), and some additional proofs were included to make the work more accessible to artists.

The interest in perspective in England is also seen in the fairly rapid publication of translations of some Italian works, notably Andrea Pozzo's *Perspective for painters and architects* (*Perspectiva pictorum et architectorum*, in Latin and in Italian, two parts, Rome, 1693, 1700). This treatise is a folio, and is lavishly illustrated. It in fact consists of a series of plates, each accompanied by about half a page of explanation. Some of the plates are taken from Pozzo's designs for illusionistic ceiling paintings. These include a spectacular rendering of a massive dome which had appeared *in situ*, as it were, in the church of S. Ignazio in Rome in 1685 (see Fig. 9.17). From the right position, the illusionism is extremely impressive—which is to say that a photograph shows what appears to be a real dome—but as one approaches or strays away from this position the huge and apparently heavy marble structures become distorted. Their apparent flexure is a decidedly bizarre visual experience. Presumably contemporaries regarded this as a fair aesthetic price to pay for the illusion. Or did they actually like the distortions? Pozzo himself argued that the distortions showed how precisely his skill had been applied in showing the correct view. In any case, there are several later versions of the dome, and they all behave in the same way. Thus a little judicious sightseeing—for instance in

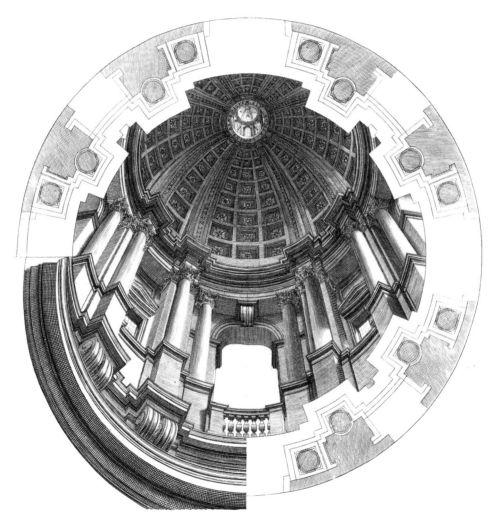

**Fig. 9.17** Plate from Andrea Pozzo, *Perspective*, showing the simulated dome painted on part of the ceiling of the church of Sant' Ignazio, Rome.

the Badia (Abbey church) at Arezzo—can go far to restore one's faith in the good sense of Le Brun and the anti-mathematical faction in the Parisian Academy of Painting and Sculpture.

Pozzo's treatise contains numerous pictures of architecture in the late seventeenth-century Roman version of Renaissance classical style. It is therefore not very surprising to find that the preface to the elegant folio English edition (London, 1707) speaks eloquently of the usefulness of the work in making the advanced style of continental architecture known in England, where it is to be hoped that native practice will be greatly improved by study of the examples it offers. This reads like the normal puffery encouraged by publishers. However, one becomes inclined to take it rather more seriously when, on turning the page, the signatories to a brief 'Approbation of the Edition' are discovered to be Christopher Wren (whom the list of subscribers describes as 'Knight, Surveyor General of Her Majesties Works'), John Vanbrugh (listed as 'Comptroller of Her Majesties Works etc.') and Nicholas Hawksmoor ('Clerk of Her Majesties Works'). With Inigo Jones (1573–1652), Wren, Hawksmoor (1661–1736) and Vanbrugh (1664–1726) are certainly the most talented architects Britain has ever produced (see Fig. 9.18)—present generation excepted, for the sake of politeness. It is clear that even professionals considered that such heavily illustrated treatises fulfilled a serious purpose. Part of Pozzo's work is certainly practical, since it deals with designs for stage sets. The list of subscribers to the edition in fact includes sundry dukes and earls (whose names are given in capitals), a scattering of notables whose names appear in Roman (Wren is in this category), while other names (given in italic) include those of four men described as 'Carpenters', one 'Master Joyner' (perhaps from Wren's staff?), one surgeon, three teachers of mathematics, one representative of the 'Mathematical Society, Jones Coffee House, Finch-lane', and a large number of painters and teachers of drawing. Most names are not followed by any professional designation, so are presumably those of people of private means. The work was clearly considered interesting and attractive by a wide variety of persons.

Treatises more or less like Pozzo's had been written throughout the sixteenth and seventeenth centuries. They continued to be written in the following centuries, though illustrations tended to become less lavish. Less luxurious texts also continued to appear. What all have in common is that they are of minimal interest to professional mathematicians of their time, because the mathematics they contain is far too elementary. Moreover, the study of perspective in its useful form seems to have gone about as far as it needs to go from the point of view of artists. The example of England appears to be typical in that almost all later works are (or claim to be) based on, extracted from, perfected out of, or otherwise derived from, some earlier text. In England the earlier text is usually Brook Taylor's. In 1754, just such a treatise was published by a friend of William Hogarth (1696–1764), and Hogarth supplied a comic frontispiece showing the perils of getting perspective wrong (Fig. 9.19).

What happens with the practical texts on perspective addressed to artists is mirrored in other texts of practical mathematics. These too are now written by authors whose mathematical skills are like those of professional mathematicians but merely on a lower level or used to a lesser extent. The formerly separate tradition of practical mathematics has essentially become a junior branch of the learned tradition. In this context Philippe de La Hire can be seen as rather old-fashioned in having such strong links to the crafts of the engineer and the painter. As the eighteenth century wore on, his learned mathematics also became old-fashioned because of its almost exclusive concern with pure geometry. Such concern was natural in a painter, but it was seen to be out of line with the thinking of most professional mathematicians. The learned tradition had absorbed

(a)

(b)

**Fig. 9.18 (a, b)** Nicholas Hawksmoor, St Anne, Limehouse. Hawksmoor's designs are notable for giving a strong sense of interlocking volumes. The lines in which surfaces cut one another are clearly articulated.

much, in both algebra and geometry, from the work of the practical mathematicians of the Renaissance. It now slowly absorbed the tradition itself. The pattern of professional knowledge came to put artists, craftsmen and mathematicians in separate social categories.

The fragmentation has proceeded to develop, and to affect the practice of both art and mathematics in ways that do not concern us here. It would, however, be unjust to abandon the story of the interrelations of mathematics and art as if the fate of projective geometry were the only matter to be considered. This book has concentrated upon geometry because the story of the rise of algebra through the practical tradition has already received so much attention from other historians. But to concentrate upon geometry, even for the good reasons that it obviously has a closer connection with the visual arts and that projective geometry eventually turned out to be important, is to run the risk of seeming to write an elegy for the demise of a particular kind of mathematics. There looms again the spectre of the sentimental notion of Renaissance Man (very rarely Woman) whose wonderfully universal knowledge gave rise to works beyond the reach of the petty

*Frontispiece.*

**Fig. 9.19** William Hogarth, *Perspective absurdities*, engraved frontispiece to John Joshua Kirby, *Dr Brook Taylor's method of perspective made easy in both theory and practice*, Ispwich, 1754.

specialists of the present day. Even in subtler forms, any such notion is rubbish. There is now more to know. Most twelve year olds in the Western world have more information about almost anything scientific than Dante did. The mathematics of the abacus schools looks familiar from schooldays, but in the Renaissance that represented the height of the learning of a lucky few. Now it is the basic material everyone has to learn, in Western Europe at least. It has continued to become more and more necessary to know some mathematics. The professional fragmentation is a result of growth, and a witness to the flourishing of an intellectual tradition that can be seen to have some of its roots in the Renaissance. Perhaps it is, in any case, too soon to see the fragmentation as definitive: with the advent of relatively cheap computers using colour visual display units and sophisticated software, mathematics and art now show signs of getting together again.

Computers are also taking us back to another problem area that was familiar to Renaissance mathematicians and natural philosophers. The use of small personal computers is leading mathematicians to take an interest in problems that are essentially those of approximation. The study of 'chaos' is the study of small variations in starting data leading to huge differences in results. We are up against a new kind of question about the use of mathematics in solving problems that involve complicated physical systems (and weather forecasting is confirmed as being an art). The situation has its parallel in the Renaissance (when weather forecasting was a task for astrologers). There was much attention paid then to issues of the applicability of mathematics in modelling the visibly untidy behaviour of the 'natural world', that is things in the 'sphere of the elements' below the Moon. The example of practical mathematics, with its habit of providing usable if not quite exact or philosophically watertight answers, seems to have played a part in resolving the matter. Philosophers were welcome to argue about the nature of number. Indeed, in 1619 Johannes Kepler, not usually one to object to the new, even argued against algebraists' numerical solutions to geometrical problems being regarded as solutions. On the other hand Galileo, as so often, spoke for the eventual victors when he maintained that one should follow the example of merchants, who knew that real ducats behaved in the same way as those in the worked examples in abacus books. That, however, is another story, belonging to the history of mathematics and science not the history of mathematics and art.

# APPENDIX: THE ABACISTS' PET TRIANGLE, WITH SIDES 13, 14, 15

The 13, 14, 15 triangle is found in Leonardo of Pisa's *Practice of geometry* (early thirteenth century). There is no reason to suppose Leonardo invented it, but since much of his mathematics comes from Islamic sources his use of the 13, 14, 15 triangle may mark its first appearance in the Latin West.

The triangle is shown in Fig. A1. It looks perfectly suitable as a non-special triangle. However, the 13, 14, 15 triangle turns out to have some very simple numerical properties. First, the height above the side 14 is 12, and second the radius of the incircle (that is the largest circle that will just fit inside the triangle, touching all three of its sides) is 4. Authors of abacus books found both these properties very helpful in designing problems with rather simple numerical solutions. For instance, since the area of a triangle is half the base times the height, the 13, 14, 15 triangle also has a simple area, namely 84.

The diagrams in Fig. A2 show how the 13, 14, 15 triangle comes by its simple height: the triangle can be dissected into two right-angled triangles, one of which (on the right in our Fig. A2(b)) is a well known whole number right-angled triangle, with sides 5, 12, 13, while the other is the even better known 3, 4, 5 triangle, faintly disguised by having all sides multiplied by 3, to give 9, 12, 15. The trick is so neat and so simple that it may be very old indeed. Since the Babylonians showed an interest in right-angled triangles, and surviving cuneiform texts deal with problems very much in the abacus book way, by means of numerical examples, I am inclined to predict that an expert on cuneiform will eventually find a Babylonian example of the 13, 14, 15 triangle. On the whole, what survives from the Greek world is rather higher level mathematics, explained in abstract terms, so if the origin of the 13, 14, 15 triangle was in a Greek schoolroom all direct trace of it has probably been lost.

In standard abacus book style, Piero della Francesca uses the 13, 14, 15 triangle to present a method that will find the height of any triangle:

But when they [the sides] are not equal, as happens when there is a triangle with *AB* 15, *BC* 14, *AC* 13 and the base is 14, multiply it into itself, [which] makes 196, multiply *AB* which is 15 into itself [which] makes

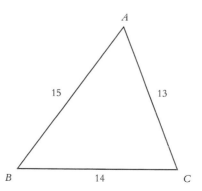

**Fig. A1**   The 13, 14, 15 triangle.

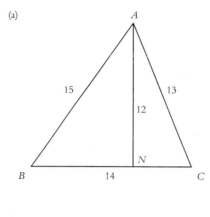

(a)

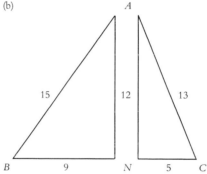

(b)

**Fig. A2**  13, 14, 15 triangle: (a) whole, (b) dissected into two right-angled triangles.

225, join it with 196 [which] gives 421; multiply *AC* which is 13 into itself [which] makes 169, take it from 421 there remains 252; divide by double the base *BC* which is 14 [so doubled] it will be 28, the result is 9: and 9 is [the distance] from *B* to the point at which the height falls [that is the foot of the perpendicular to the base]. Multiply 9 into itself [which] makes 81, and multiply *AB* which is 15 into itself [which] makes 225; subtract 81 there remains 144: its root is the height, which is 12. And this method can be used for all triangles.

This method depends upon using Pythagoras' Theorem (Euclid, *Elements*, Book 1, Proposition 47), in triangles *ANB* and *ANC*. For the sake of clarity, and concision, we shall employ modern algebraic notation, giving symbols to the various lengths as shown in Fig. A3.

First consider triangle *ANB*. *AN* is the height of triangle *ABC*, so *ANB* has a right angle at *N*, and Pythagoras' Theorem gives

$$x^2 + h^2 = c^2. \tag{1}$$

Since we have set *BN* = *x* and the total length of *BC* is *a*, the length *NC* is *a* − *x*, and Pythagoras' Theorem in triangle *ANC* gives

$$(a - x)^2 + h^2 = b^2. \tag{2}$$

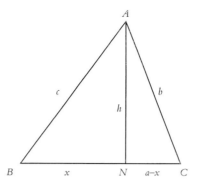

**Fig. A3**   General triangle with height to illustrate Piero's formula.

To eliminate $h^2$ we subtract equation (2) from (1), which, using the binomial expansion for $(a-x)^2$ (deducible from Euclid), gives

$$x^2 - (x^2 - 2ax + a^2) = c^2 - b^2.$$

This tidies up into

$$2ax = a^2 + c^2 - b^2,$$

which gives us the formula that Piero quoted, namely

$$x = (a^2 + c^2 - b^2)/2a,$$

though he put it in the numerical form

$$BN = (14^2 + 15^2 - 13^2)/(2 \times 14).$$

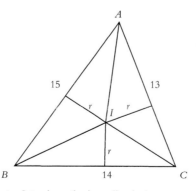

**Fig. A4**   13, 14, 15 triangle with in-centre *I*, to show the in-radius is 4.

Piero's formula for the height, $AN$ (= $h$), depends upon the use of Pythagoras' Theorem in triangle $ABC$. This gives us equation (1) above. Rearranging the terms gives us Piero's result, namely

$$h^2 = c^2 - x^2.$$

That is, in numerical terms

$$h = \sqrt{(15^2 - 9^2)}.$$

The radius of the incircle of the 13, 14, 15 triangle can be found by dissecting the triangle into three triangles with a common vertex at the centre of the incircle, as shown in Fig. A4. If we now consider the area of the complete triangle as the sum of those of the three smaller ones we obtain $r$ = 4. Abacus books essentially use this same method.

# BIBLIOGRAPHY

Cappel, C. B. (1997). *Design theory and workshop practice*. Cambridge University Press.

Chaboud, M. (1994). Desargues Lyonnais. In *Desargues en son temps* (ed. J. Dhombres and J. Sakarovitch). Blanchard, Paris, pp. 433–52.

Davis, M. D. (1977). *Piero della Francesca's mathematical treatises: The 'Trattato d'abaco' and 'Libellus de quinque corporibus regularibus'*. A. Longo Editore, Ravenna.

Euclid, *The Thirteen Books of Euclid's Elements* (ed. and trans. T. L. Heath). 1st edn Cambridge University Press, 1908; 2nd edition, Dover Press, New York, 1956.

Field, J. V. (1985). Giovanni Battista Benedetti on the mathematics of linear perspective. *Journal of the Warburg and Courtauld Institutes*, **48**, 71–99.

Field, J. V. (1986). Piero della Francesca's treatment of edge distortion. *Journal of the Warburg and Courtauld Institutes*, **49**, 66–99 and Plate 21c.

Field, J. V. (1987). Linear perspective and the projective geometry of Girard Desargues. *Nuncius*, **2.2**, 3–40.

Field, J. V. (1988) Perspective and the mathematicians: Alberti to Desargues. In *Mathematics from manuscript to print* (ed. C. Hay). Oxford University Press, pp. 236–63.

Field, J. V. (1993). Mathematics and the craft of painting: Piero della Francesca and perspective. In *Renaissance and revolution: humanists, craftsmen and natural philosophers in early modern Europe* (ed. J. V. Field and F. A. J. L. James). Cambridge University Press, pp. 73–95.

Field, J. V. (1995a). A mathematician's art. In *Piero della Francesca and his legacy* (ed. M. A. Lavin). (Studies in the History of Art, no 48, Center for Advanced Study in the Visual Arts, Symposium Papers XXVIII) pp. 177–97. Washington, DC, National Gallery of Art.

Field, J. V. (1995b). Piero della Francesca and the 'distance point method' of perspective construction. *Nuncius*, **10.2**, 509–30.

Field, J. V. and Gray, J. J. (1987). *The geometrical work of Girard Desargues*. Springer Verlag, New York.

Field, J. V., Lunardi, R. and Settle, T. B. (1988). The perspective scheme of Masaccio's *Trinity* fresco. *Nuncius*, **4.2**, 31–118.

Franci, R. and Toti Rigatelli, L. (1985). Towards a history of algebra from Leonardo of Pisa to Luca Pacioli. *Janus*, **72**, 17–82.

Goldthwaite, R. (1980). *The building of Renaissance Florence*. The Johns Hopkins University Press, Baltimore.

Goldthwaite, R. (1993). *Wealth and the demand for art in Italy 1300–1600*. The Johns Hopkins University Press, Baltimore.

Gray, J. J. (1979). *Ideas of space: Euclidean, non-Euclidean and relativistic*. Oxford University Press.

Hall, A. R. (1983). *The revolution in science, 1500 to 1750*. Longman London. (First published as *The scientific revolution, 1500 to 1800*. London, 1954.)

Hall, A. R. (1992). *Isaac Newton, adventurer in thought*. Blackwell, Oxford.

Hollingsworth, M. (1984) The architect in fifteenth-century Florence. *Art History*, **7**(4), 385–410.

Holmes, G. (1986). *Florence, Rome and the origins of the Renaissance*. Oxford University Press.

Jaywardene, S. A. (1976). The *Trattato d'abaco* of Piero della Francesca. In *Cultural aspects of the Italian Renaissance: essays in honour of Paul Oskar Kristeller*. Manchester University Press, pp. 229–43.

Kemp, M. J. (1984). Construction and cunning: the perspective of the Edinburgh Saenredam. In *Dutch church painters: Saenredam's 'Great Church at Haarlem' in context* [Exhibition catalogue] (ed. H. Macandrew). National Gallery of Scotland, Edinburgh.

Kemp, M. J. (1990). *The science of art: optical themes in western art from Brunelleschi to Seurat*. Yale University Press, New Haven.

Kemp, M. J. (1995). Piero and the idiots. In *Piero della Francesca and his legacy* (ed. M. A. Lavin). (Studies in the History of Art, no 48, center for Advanced study in the Visual Arts, Symposium Papers xxviii) pp. 199–211.

Lavin, M. A. (1972). *Piero della Francesca. The Flagellation of Christ*, New York. (Reprint, with additional bibliography, University of Chicago Press, 1990.)

Lindberg, D. C. (1976). *Theories of vision from al-Kindi to Kepler*. University of Chicago Press.

Macandrew, H. (ed.) (1984) *Dutch church painters: Saenredam's 'Great Church at Haarlem' in context* [Exhibition catalogue]. National Gallery of Scotland, Edinburgh.

Mesnard, J. (1994). Desargues et Pascal. In *Desargues en son temps* (ed. J. Dhombres and J. Sakarovitch). Blanchard, Paris, pp. 87–99.

Parker, R. H. and Yamey, B. S. (eds) (1994). *Accounting history: some British contributions*. Oxford University Press.

Picon, A. (1994). Girard Desargues ingénieur. In *Desargues en son temps* (ed. J. Dhombres and J. Sakarovitch). Blanchard, Paris, pp. 413–22.

Pirenne, M. H. (1970). *Optics, painting and photography*. Cambridge University Press.

White, J. (1993). *Art and architecture in Italy 1250–1400*, 3rd edn. Yale University Press, New Haven.

Wittkower, R. and Carter, B. A. R. (1953). The perspective of Piero della Francesca's 'Flagellation'. *Journal of the Warburg and Courtauld Institutes*, **16**, 292–302.

Wright, L. (1983). *Perspective in perspective*. Routledge and Kegan Paul, London.

Zeuthen, H. G. (1903). *Geschichte der Mathematik in XVI. und XVII. Jahrhundert* (ed. B. Meyer). B. G. Teubner Verlag, Leipzig.

# Index

Bold numbers denote reference to illustrations